EVERYTHING YOU ALWAYS WANTED TO KNOW ABOUT STATISTICS, BUT DIDN'T KNOW HOW TO ASK

Second Edition

EVERYTHING YOU ALWAYS WANTED TO KNOW ABOUT STATISTICS, BUT DIDN'T KNOW HOW TO ASK

Second Edition

James K. Brewer
Florida State University

KENDALL/HUNT PUBLISHING COMPANY
4050 Westmark Drive   Dubuque, Iowa 52002

Copyright © 1978, 1996 by Kendall/Hunt Publishing Company

Library of Congress Catalog Card Number: 96-76020

ISBN 0-7872-2162-7

All rights reserved. No part of this publication may be reproduced, stored in a retrieval system, or transmitted in any form or by any means, electronic, mechanical, photocopying, recording, or otherwise, without the prior written permission of the copyright owner.

Printed in the United States of America

10   9   8   7   6   5   4   3   2

INTRODUCTION

The purpose of this text is to provide the mathematically naive reader with a common-sense question and answer approach to the essentials of basic statistics. The questions are typical of those asked by beginning students which the author has taught in all areas of behavioral science, engineering and business. The answers are intended to guide the reader to a sound understanding of the concepts of descriptive and inferential statistics rather than emphasizing "number crunching."

The text may be used as the basic instructional text for undergraduate or elementary graduate statistics if accompanied by elucidating lectures from a competent instructor. It has been found to be equally useful as a refresher or supplemental source for the once-initiated who have been away from mathematics and statistics for some time as well as those who need a readable source to obtain the ideas, notation and expressions of statistics.

Doubtless, there are many approaches to answering the questions in this material as well as additional questions which should be asked and answered. The reader is invited to question the interpretations and answers given and to utilize any information source at his disposal to reach a satisfactory level of interpretation and understanding. In doing so the reader should keep in mind that there is almost always more than one interpretation for any answer or set of observable data. The intent of the previous statements, as well as the underlying philosophy of this material, is expressed very well by F.R. Moulton:

> ... every set of phenomena can be interpreted consistently in various ways. It is our privilege to choose among the possible interpretations the ones that appear to us most satisfactory, whatever may be the reasons for our choice. If scientists would remember that various equally consistent interpretations of every set of observational data can be made, they would be much less dogmatic than they often are, and their beliefs in a possible ultimate finality of scientific theories would vanish.

CONTENTS

List of Tables, **ix**
General Terminology, **1**
Measures of Central Location, **6**
Measures of Dispersion, **11**
Measures of Association, **20**
Measures of Predictability, **28**
General Inference Concepts, **34**
Specific Hypothesis Tests, **46**
Bayesian Statistics, **64**
Some Statistical Reminders, **66**
Notation, **67**
Appendix A: Basic Mathematics Skills Refresher, **77**
  Summation Notation, **87**
Appendix B: Finding the Square Root, **96**
Appendix C: Probability, **100**
Appendix D: Normal Probability Distribution, **104**
Appendix E: Example Calculating Formula for Minimum Sample Size, **112**
Appendix F: Myths and Misconceptions in Behavioral Statistical Methodology, **113**
References, **119**
Index, **121**

LIST OF TABLES

Table A: t-table, **69**
Table B: F-table, **70**
Table C: Chi-Square Table, **74**
Table D: Normal Curve Table, **75**

GENERAL TERMINOLOGY

What is statistics?

Statistics is a branch of applied probability theory.

That's impressive—but what is the purpose of statistics?

There are two basic purposes. The first is to describe data quantitatively, and the second is to make inferences with certain degrees of uncertainty. The uncertainty is in the form of a probability or chance of making specific types of errors relative to the inference.

What is "data"?

Data is the numerical information collected from a sample.

What is a sample?

A *sample* is a subgroup of a population.

What is a population?

A *population* is the entire group, set of scores or observations, concerning which you wish to make an inference or a descriptive statement. A population is almost always assumed to be infinite in size.

Can you give me an example which ties these terms together?

Be glad to. Suppose for some reason you were interested in I.Q. scores of 12 year-old boys. I.Q. scores of all the 12 year-old boys, which could be or have been collected, would be an example of a *population*. The set of I.Q. scores which come from Ms. Jones' 12 year-old boys would be a *sample* of I.Q. scores. The *data* would be the numerical information, I.Q. scores, in this sample. (Note that a sample could refer to a set of people also; e.g., Ms. Jones' 12 year-old boys, but generally samples are taken of the people primarily to obtain data; i.e., scores from them.)

Another term which you will need is *variable*; i.e., the property whereby one observation differs from another. The variable in this example is I.Q. There are two basic types of variables: *continuous* and *discrete*. A *continuous* variable is one which takes on all values within a defined range. That is, it has no gaps or breaks in it. For example, the variable temperature is continuous since the mercury must move through all intermediate points between any two degree markings. It does not jump across certain degree markings. A *discrete* variable is one which is not continuous and therefore has breaks or gaps. For example, the variable sex is discrete since, between male and female, the variable cannot take on any other values due to the fact that there are no others (barring the odd case, of course).

Note that your measuring instrument will always produce discrete scores or numbers even though the variable is continuous. This is because there are no physical instruments which can measure all possible points between two other points. For example, you could believe that attitude is continuous; i.e., people could have every attitude value between "good" and "bad." If you denoted "good" with a "1" and "bad" with a "5," there is no instrument which will pick up every number between 1 and 5, since there are an uncountably infinite number of points between 1 and 5.

Is "probability" what I think it is?

It probably is. Forgive the redundancy, but almost everyone at your stage in life has some notion of probability, chance, likelihood, possibilities, etc., and almost all these notions function equally well. A brief discussion of the several aspects of probability, along with some basic rules used in this material, are given in Appendix C, Page 100. It would be wise to read through Appendix C at this time—if for nothing more than a refresher.

What is meant by "types of error relative to inference"?

Since inferences are made in regard *to* a population *from* a limited sample, there will be a possibility (probability) of making errors; i.e., in being wrong relative to the inference drawn. The specific types of errors and the probabilities associated with them will be discussed in detail when specific statistical inferences are presented later. Can you wait until then?

I'll try.

Thank you.

What is the distinction between description and inference?

If one desires to do no more than quantitatively describe some characteristics of the data at hand and say nothing beyond the data itself, then this is the *descriptive* aspect of statistics. Examples are the calculations of some statistics, such as; arithmetic means, proportion of students who like statistics, etc., for a set of data and saying things like, "The mean I.Q. of Ms. Jones' 12 year-old boys is 105," or, "The proportion of students in this class who like statistics is .21."

If one desires to make a statement beyond the data in hand, then this is the *inferential* part of statistics. For example, if one wishes to infer from the sampled data, with some degree of uncertainty, that the parameter, population mean, is not equal to 105, then an inference to a population (whence cometh the sample) is desired.

You referred to the arithmetic mean as a "statistic." Is a "statistic" used with samples or a population?

Samples. Any numerical characteristic describing a sample is called a *statistic*. Correspondingly, a numerical characteristic describing a population is called a *parameter*. For example, the true proportion of all students who like statistics is a parameter, and a statistic would be the proportion in a particular classroom who like statistics. Other statistics and parameters will be encountered later, so the distinction must be clear at this point.

What if I have the whole population in hand? *[handwritten: what about those #'s @ other times or places]*

Obviously, if there is nothing beyond your data to which an inference can be made, then you can make no inference. Recall here that population refers to a population of numbers (scores, measures, etc.). If you have obtained your numbers by taking measurements on a set of objects (or people), then what you have is a sample of such measurements even though there are no more objects (or people) to be measured. This is so, since other numbers could occur if other measurements were made at other times and places on the same objects, i.e., measurement error is possible.

What do you mean by measurement?

Measurement is the process of assigning a number to an object, place or person. The assignment may come about by way of an instrument, formula, observation, or test. An example of measurement would be the assigning of I.Q. scores (via some instrument) to the 12 year-old boys in Ms. Jones' room. The single score on each boy is a measurement *observation*.

What kinds of measurement are there?

There are four basic kinds of measurement. They are called nominal, ordinal, interval and ratio.

1. *Nominal measurement:* The assignment of numbers to objects so that the numbers do no more than identify the objects. For example, football jersey numbers.

2. *Ordinal measurement*: The assignment of numbers to objects so that the numbers imply order as to magnitude, importance, etc. For example, assigning a "1" to the tallest person in the room, a "2" to the next tallest, etc.

3. *Interval measurement*: The assignment of numbers to objects so that the numbers are ordinal and intervals are equal, that is, a change from 50 to 60 is the same as from 60 to 70. For example, the marks on a thermometer are interval.

4. *Ratio measurement*: The assignment of numbers to objects so that the numbers are interval and there exists an absolute zero. For example, height and weight are ratio scale measurements.

Which of these types of measurement will I be concerned with in this material?

Primarily, *interval* since it will be a fundamental assumption of the statistical theory with which most inferences will be made. It will be made clear to you when it is an assumption and when it is not. Bear in mind, one cannot look at a set of numbers and tell what their measurement scale is. Only by knowing how and why the scores were collected can a judgment be made.

O.K. but what if there can be no more measurements made on a group of people?

If there are conceivably no more measurements which can be made, then you have the population and no inferences should be made. For example, if you were interested in how many people there were in the class at a specified time, then a simple count would suffice and there is no inference to be made. The population here consists of that one number representing the number of people in class.

How often can I expect to have a population of numbers in hand?

Almost never. With almost any situation it is easy to conjure up a population of numbers very much larger than the set of numbers in hand which could have been selected. If you can obtain a population, do so. Never take a sample if a population can be readily obtained, since a sample will only be a subpart of a population.

Does this mean that I will, therefore, have to make inferences only?

Not necessarily. There's nothing wrong with calculating statistics on sample data for descriptive purposes. It is, however, for the primary purpose of inference that samples are obtained.

What ways can I describe data from a sample?

There are several different ways to describe data from a sample and some of the more common are called:

(a) Measures of central location — mean, median, mode
(b) Measures of dispersion
(c) Measures of association
(d) Measures of predictability

Each of these will be discussed in order.

MEASURES OF CENTRAL LOCATION

These measures deal with calculating an index or number to describe the central value of the data, that is, a value which describes the general location of the data at hand. It may or may not be representative of the data, but is generally used as such.

There are many such measures but the most common is the one you already know, namely, the *arithmetic average* or *mean* which consists simply of the sum of all the scores divided by the number of scores. Symbolically, the *arithmetic average* of a sample of size n is

$$\overline{X} = \sum_{i=1}^{n} X_i/n = \frac{X_1 + X_2 + \ldots + X_n}{n},$$

where X is the variable of interest and X_i represents the *ith* score on that variable.

What's that funny-looking symbol "Σ"?

That's a Greek letter (upper case s). It is a symbol with which you must be familiar and the latter part of Appendix A should provide you with this familiarization. It is a shortcut technique which you will need from this point on.

Does $\overline{X}$ estimate the true population average?

Yes, and the symbol used for any population average is μ (Greek lower case m). To elaborate further on this statistic $\overline{X}$ (a "statistic," you will recall, being any numerical value based on a sample) except to say that it has "nice" mathematical properties is a waste of your time, since you know how to find it and have known for years.

Will you work out an example with $\overline{X}$?

Surely. The numbers will be kept quite simple, since the concept of $\overline{X}$ is the most important thing, not "number crunching."

Suppose you have five scores resulting from the administration of an attitude scale and they are 3, 5, 4, 2, 6.

The average, $\overline{X}$, of this sample of scores is

$$\sum_{i=1}^{n} \frac{X_i}{n} = \sum_{i=1}^{5} X_i/5 = \frac{3 + 5 + 4 + 2 + 6}{5} = \frac{20}{5} = 4.$$

Quite often scores are shown in grouped form, for example:

Score	Frequency
0-2	3
3-5	5
6-8	4
9-11	2

What is this all about?

This is a simple but approximate way to describe the set of 14 scores. Notice that you do not know what each score is; only how many (frequency) scores there are in each interval (range of scores). In the score range 3 to 5 you know that there are five scores but you do not know the exact value of each score, where they are located in the interval, or if they are all equal. (If you knew the value of each score, $\overline{X}$ could be calculated directly.)

Any presentation of the scores with their frequencies is called a *frequency distribution*.

If I do not know each score, how can I find $\overline{X}$?

You don't find $\overline{X}$, but you can find a reasonable approximation of it.

What do you mean "reasonable"?

This means an approximation that makes sense under the conditions and is a defensible alternative.

O.K., what do I do to approximate $\overline{X}$ in a reasonable way?

Since you do not know the exact location of each score in the intervals shown you could assume that they are all located at the midpoint of each interval. This would appear to be the simplest of assumptions to make and certainly simplifies the arithmetic. Is this a reasonable assumption?

It seems O.K., but how do I find the mean if I pretend the scores are all at the midpoint of each interval?

Let's assume that each score in a given interval is located at the midpoint of the interval. Now the scores and the frequency of each become

Midpoint Score (X_i)	Frequency (f_i)
1	3
4	5
7	4
10	2

Since you are now pretending that there are three 1's, five 4's, four 7's, and two 10's, the mean can be calculated as

$$\sum_{i=1}^{4} \frac{X_i f_i}{14} = \frac{(1)(3) + (4)(5) + (7)(4) + (10)(2)}{14}$$

$$= \frac{71}{14} = 5.1,$$

(Note that the above sum runs from 1 to 4 since there are only four intervals.)

Does this work for any set of intervals?

Yes, regardless of the form of the intervals, one need only multiply each midpoint by the frequency, add these products and then divide by the total frequency, i.e., n.

If I knew each score could I form my own intervals and calculate $\overline{X}$ from this set of intervals?

Yes you could, but only if your sample size was so "humongous" that the calculation could not be easily made from the *raw data* formula

$$\overline{X} = \sum_{i=1}^{n} \frac{X_i}{n}.$$

Recall also, that a grouped data answer for $\overline{X}$ is an approximation, not necessarily the exact value of the sample mean. For your purposes, however, this approximation using grouped data may be sufficient.

May I have a raw data and a grouped data problem to check my understanding of $\bar{X}$?

Certainly, you may.

Problem 1. Find $\bar{X}$ for the sample set of raw data $-1, 0, 2, 3, 3, 5$.

Problem 2. Find $\bar{X}$ for the following grouped data (frequency distribution).

Scores (X_i)	Frequency (f_i)
10-14	2
15-19	3
20-24	5
25-29	3
30-34	2

Answers: (1) $\bar{X} = 2$
(2) $\bar{X} = 22$

Aren't there other measures of central location?

Yes, but they don't possess the "nice" mathematical properties of $\overline{X}$. The two most common are:

(a) *Median:* A value such that half the scores are above it and half are below it. There's no formula needed here. It's a matter of looking at the data and finding it. For example, suppose your sample of scores is

$$1, 2, 3, 3, 5, 7, 9, 9.$$

Now the median is *somewhere* between 3 and 5. It could be anywhere but you may calculate a *convenient* one by taking the average of 3 and 5 to get 4.

(b) *Mode:* The score which occurs most frequently in a sample of scores. In the above example both 3 and 9 are modes. This isn't a very interesting measure of central location, so no more time will be spent on it.

Are there grouped data methods for the median and mode?

Yes, but they are so little used that it's not worth your time to discuss them here. Most statistics texts will show you how—if you ever need the grouped data form.

You have spent more time on $\overline{X}$ than on the other measures of central location. Why?

The arithmetic average, $\overline{X}$, is the primary measure of central location because it is used in the development of other descriptive statistics and is the major statistic relative to making inferences.

How will $\overline{X}$ be used to develop other descriptive statistics?

I was hoping you'd ask that question. It lets me lead you into:

MEASURES OF DISPERSION

These measures are concerned with how the scores disperse, spread out, scatter, vary, etc. There are several such indices of dispersion but only two will be discussed here—range and variance.

(a) *Range:* The difference between the largest and smallest score, ignoring the sign of the difference. It is apparent that this value gives you some idea of the extent of spread of the scores but not a very elaborate one. It is one of those descriptive statistics with which you are familiar, so nothing further need be said about it. (The range for the data on page 10 is 8.)

(b) *Variance:* This value gives an index of the extent of cluster of the scores about the central location value, $\overline{X}$. It is the most common, and seemingly complicated, of the measures of dispersion but, as you have guessed by now, it has "nice" mathematical properties. The notation for the *sample variance* will be

$$S^2 = \sum_{i=1}^{n} (X_i - \overline{X})^2/(n-1)$$

It is apparent that the form of the expression gives an indication of how far from $\overline{X}$ the scores generally are since S^2 can be large only if the differences, $(X_i-\overline{X})$, are generally large, regardless of whether the scores are above or below the mean, $\overline{X}$. S^2 estimates the population variance, denoted σ^2. The variance for the sample data 1, 2, 3, 4, 5, is $[(1-3)^2 + (2-3)^2 + (3-3)^2 + (4-3)^2 + (5-3)^2] \div 4 = (4 + 1 + 0 + 1 + 4)/4 = 10/4 = 2.5$.

Why are the differences squared?

The differences are squared primarily because if they were not the numerator of S^2 would always be zero, regardless of what the scores were. (An index of dispersion which was the same for all data would, of course, be worthless.) Without the square the numerator of S^2 would be

$$\sum_{i=1}^{n} (X_i - \overline{X}) = \sum_{i=1}^{n} X_i - n\overline{X} = \sum_{i=1}^{n} X_i - \sum_{i=1}^{n} X_i = 0,$$

which says simply that the sum of all deviations from $\overline{X}$ will be zero. So squaring each deviation seems like a logical thing to do, since it will still reflect the spread or dispersion of the scores.

In the formula for $\overline{X}$ you divided by n. For S^2 you divided by n − 1, why?

The reason for n − 1 is mathematical in that, if n were used,

$$\sum_{i=1}^{n} (X_i - X)^2 / n,$$

would not be an unbiased estimator of σ^2, the population variance. With n − 1 used instead of n, S^2 is an unbiased estimator of σ^2. (The proof of this will have to be taken on faith at this stage. Any standard statistical theory text will provide the necessary proof.)

What is an "unbiased estimator" since this seems to be crucial to the reason for using n − 1?

S^2, as given on page 11, is an unbiased estimator of σ^2 because if all S^2 values were calculated using all possible samples of size n, then the arithmetic average (mean) of all such S^2 values would be *exactly* σ^2. By the way, $\overline{X}$ is an unbiased estimator of μ for the logical reason that the mean of all sample means of size n is the population mean. Obviously, for very large samples it makes no material difference whether you use n or n − 1 in the calculation of S^2 but n − 1 is preferred for the above reason and for later inferential reasons.

What about other measures of dispersion?

There are no others of real importance statistically or practically, so S^2 will be used throughout as *the* measure of spread or dispersion.

I've heard in the past of a thing called "standard deviation." What is it?

Standard deviation (S.D.) is simply the positive square root of variance, nothing more or less. It, therefore, gives no more nor less information concerning dispersion than S^2. It is denoted, imaginatively, as S and will be used later in describing and developing inferential statistics. Notationally, sample standard deviation is

$$S = \sqrt{\sum_{i=1}^{n} (X_i - \overline{X})^2/(n-1)}.$$

As you can see it is "sort of" the average distance that each score is from $\overline{X}$ and is expressed in the unit of the measure rather than in squared units like S^2.

I think I have a handle on S^2. Give me a problem or two and see how I do (note the poetic nature of my request).

Your poetry leaves much to be desired. Let's hope your calculating ability doesn't.

Find the variance (S^2) and standard deviation for the following (complicated) set of sample raw data

$$-1, 0, 2, 3, 3, 5$$

Answer: $S^2 = 24/5 = 4.8$; $S = 2.19$

What if the data are in grouped (or frequency distribution) form?

You proceed the same way you did in calculating $\overline{X}$, namely, pretend that each score in a particular interval is located at the midpoint of the interval and use the raw data formula for S^2.

You mean that's all there is to the calculation of sample means and variances?

Exactly. With more and larger numbers the arithmetic gets tedious but that's the only difference.

It seems to me there ought to be more to $\overline{X}$ and S^2 than this. Maybe there are simply other ways to calculate S^2 in particular formulas, which avoid the mess of taking all those differences and squaring them. Are there such alternative formulas?

Yes, and they are referred to as "calculating" formulas. They are equivalent to the "definitional" formula on page 11 and may be handy if many large numbers are used. You may use them interchangeably with the definitional formula anytime you wish.

Calculating formulas for S^2

(1) $$S^2 = \frac{n \sum_{i=1}^{n} X_i^2 - (\sum_{i=1}^{n} X_i)^2}{n(n-1)},$$

or equivalently,

(2) $$S^2 = \frac{\sum_{i=1}^{n} X_i^2 - n\bar{X}^2}{n-1}.$$

Some rules on $\bar{X}$ and S^2

1. If a constant is added to (or subtracted from) each score, then $\bar{X}$ is altered by that amount but S^2 is unaltered. For example, if five is added to each score, then the new mean is $\bar{X} + 5$ and S^2 is unchanged.

2. If a constant is multiplied by (or divided into) each score, then $\bar{X}$ is altered by the amount of that constant and the variance is altered by the square of that amount. For example, if five is multiplied by each score, then the new mean is $5 \cdot \bar{X}$ and the new variance is $(25)S^2$. (Note that the new S.D. is $5 \cdot S$ and would be $5 \cdot S$ even if each score had been multiplied by a negative 5.)

You mean I could convert a set of scores to have any $\bar{X}$ and S^2 I wanted simply by adding, subtracting, multiplying or dividing each score by numbers chosen by me?

Yes, and this whole process is referred to as *score transformation* or *coding*. The most common such transformation results in the generation of a set of scores which have a $\bar{X}$ of 0 and a S^2 of 1. These scores are called *standard unit scores*. Do you see how to convert a set of scores to standard unit scores?

I think so. Would you subtract $\bar{X}$ from each score and then divide this set of transformed scores by S?

Exactly. By subtracting $\bar{X}$ from each score the new mean of the *deviation* scores is $\bar{X} - \bar{X} = 0$. Notice that subtracting $\bar{X}$ from each score does not affect its S^2 or S. Now, dividing each of the deviation scores by S makes the transformed scores have a mean of $0/S = 0$ and a variance of $S^2/S^2 = 1$. How about that, fans?

That's pretty slick, but why transform scores to standard unit scores?

Besides the fact that the transformed scores will be considerably smaller in numerical value, this standard unit score form is crucial to working with special mathematical distributions which are needed in making inferences. Basically, it simplifies both the theory and the application of statistics.

As an exercise and review of $\overline{X}$ and S see if you can convert the following scores to standard unit scores.

$$4, 4, 6, 8, 8.$$

Answer: $-1, -1, 0, 1, 1$

Are there other kinds of standard scores?

Yes, there are several which have specific means and standard deviations. For example, Graduate Record Exam (GRE) scores are standard scores with a mean of 500 and a standard deviation of 100. In fact, almost all "standardized" test scores are simply transformed raw scores.

Notational note: Generally the raw scores will be designated by upper case letters, e.g., X, and standard scores will be designated by lower case letters, e.g., x. Specifically, the mathematical relationship between raw (X scores) and standard unit scores (z scores) is usually expressed

$$z = \frac{X - \overline{X}}{S}$$

I have often seen data presented in graphical form, with lines, charts, etc. What is this all about?

This is simply another way to describe data. The graphical methods used are many and varied in format. There are, however, a few which are common enough in statistical work to warrant mentioning here, e.g., a *histogram*, a *frequency curve* and a *density curve*. An illustration is the best way to describe what these represent.

Suppose a set of 20 scores is given in frequency distribution form (or have been put in this form by you). Let the distribution be

X	f
1-10	2
11-20	5
21-30	7
31-40	4
41-50	2

Now if you let the vertical axis represent the frequency of scores (f) and the horizontal axis represent the score value (X), the above scores can be represented in block form as shown in Figure 1.

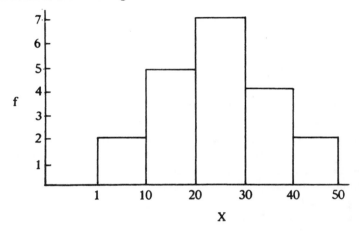

Figure 1. Frequency histogram.

This is called a *histogram* or *frequency histogram*. If each frequency is divided by 20, the sample size, the picture is called a *probability* or *relative frequency* histogram. Note in this latter case the height of each block is a probability or proportion and the sum of these heights (probabilities) is one. (Check this for yourself.)

If a dot is centered at the top of each of the blocks in the above picture and all these dots are joined with a smooth curve, Figure 2 will result.

This graphical representation of the same data is called a *frequency curve* (or frequency distribution curve). If each frequency is divided by 20, the result is called a *density curve* (or relative frequency curve).

In most of statistical inference work, the density curve is the most common way of describing the shape and characteristics of the underlying, or true, form of the population of data which produced the sample being used. Several specific density curves will be utilized later when inferences to populations are made.

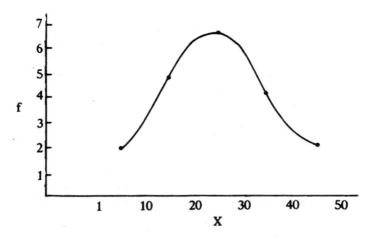

Figure 2. Frequency curve.

Can these frequency distributions take on different shapes?

Yes, and there are names to describe the general shapes of some of them. For example, a *symmetric* distribution is one in which there will be approximately the same number of small and large scores. Figure 3 shows several types of symmetric distributions.

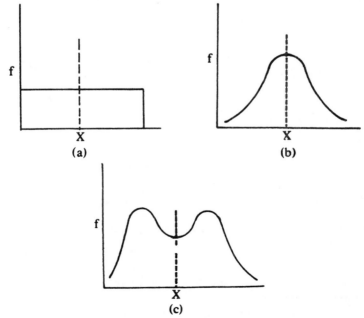

Figure 3. Symmetric distributions.

Figure 3(a) is called a *uniform* distribution (each score is equally likely to occur), Figure 3(b) is a *unimodal* (one mode) distribution and Figure 3(c) is a *bimodal* (two modes) distribution. Figure 3(b) will be the most commonly assumed symmetric distribution and as a result will be the one you will spend much time on later. Note that if each graph in Figure 3 were folded on the dotted line the two halves would match. This is the essence of symmetry.

Certainly not all distributions are symmetric. Are there terms for asymmetric distributions?

You are quite right and there are terms for some types of asymmetry. You will recall that $\overline{X}$ was a descriptive statistic to describe the central location of a distribution or representative score and that S^2 was an index of how the distribution spread about the value $\overline{X}$. There are several other descriptors of distributions and the most common one is called *skew*. Skew is one index of the asymmetry of a distribution. (Obviously, skew will be zero for symmetric distributions.)

Does it have a formula like $\overline{X}$ and S^2 ?

Yes, it does, but instead of giving you another formula to worry about, a description of what *positive* and *negative* skew would look like will be given. Figure 4 shows distributions which would yield positive and negative skew values if one bothered to calculate them.

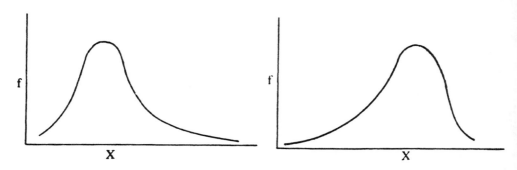

Figure 4(a). Positive skew.   **Figure 4(b).** Negative skew.

What kinds of scores would result in positive and negative skew values?

A positive skew could result if the X variable were, for example, reading scores on a test administered to children with severe reading problems. It would be reasonable in this case for there to be a relatively large number of low scores and a relatively small number of high scores. Can you change this example slightly to provide an example of a negative skew?

If we administered the same test to exceptionally good readers and looked at their reading scores, would that yield a negative skew?

Probably so, as you would most likely have a greater proportion of high scores than low scores since the sample of people used was composed of exceptional readers (You gave an excellent example).

Thank you.

MEASURES OF ASSOCIATION

What do you mean by "measures of association"?

These measures involve the assigning of a numerical value to describe the degree (strength, magnitude) of a relationship between two variables.

You have doubtless used or heard expressions like, "height and weight are correlated," "attitude is associated with performance" or "intelligence is related to achievement." What is meant by these expressions is that for some reason or other one of the variables, say height, goes up and down with the other variable, weight, thus implying a relationship between height and weight. Assigning a number to the *degree* of this relationship between two variables, via an index, is what "measures of association" is all about.

The most common general term used to describe a relationship between two variables is called *correlation* and several indices are used to quantify the strength or degree of this correlation.

What kinds of indices of correlation are there?

There are many, but the concern here will be with only two types: *Pearson Product Moment [PPM]* and the *Spearman Rho [SR]* Correlation Coefficients. They are both derived from the same basic formula but will look quite different, due to the assumptions for each of them.

Pearson Product Moment Correlation Coefficient

The PPM is the most common and widely used of the correlation coefficients and has the most restrictive assumptions. It is an index of the degree or magnitude of *linear* relationship between two variables. For example, suppose you felt that the variable performance, denoted the Y variable, was truly a linear function of the variable attitude, denoted X. Suppose further that both X and Y were variables measured on an interval scale, then the PPM would be a reasonable index of association.

What do you mean by "linear relationship" and "linear function"?

In nonmathematical terms these expressions simply mean that values of X and Y go up and down together, or in opposite directions, in straight-line fashion. For example, pretend you could measure everybody on both performance (Y) and attitude (X). Now it is obvious that for each attitude score there may be hundreds of different performance scores. Suppose that for each of these different attitude scores you averaged all the performance scores which are associated with them. It is said that the true relationship between X and Y is linear if all these averages ($\overline{Y}$ values) lie on a straight line. (Let $\overline{Y}$ and X be, respectively, the vertical and horizontal axis, then Figure 5 shows an example of a true linear relationship).

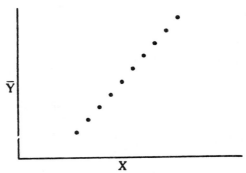

Figure 5. A true linear relationship.

Note that for a linear relationship to exist, all X and Y values *do not* necessarily lie on a straight line-only the averages of one of the variables.

What if I could not make those assumptions of linearity and interval scale?

Then the *Spearman Rho Correlation Coefficient* might be an appropriate alternative. It will be discussed after the PPM, if that's O.K.

O.K. I'll wait. Suppose I can feel comfortable with these assumptions. What now?

Now you need a formula to calculate, from a sample, a number which approximates the degree of this true linear relationship since you cannot ever know the true degree of linear relationship. The index PPM is designed to range between −1 and 1 with 1 and −1 indicating perfect linear relationships, that is, all the plotted X, Y points in the sample lie on a straight line. The −1 indicates that they all lie on a straight line but large values of X are paired with small values of Y, and vice versa. A +1 value for PPM indicates that all the sample pairs of points are on a straight line and that large values of X are paired with large values of Y and small values of X are paired with small values of Y.

The sample PPM index is denoted with the letter r and is an approximation of the true population value ρ (Greek lower case r). The raw data formula for r with n pairs of scores is

$$r = \frac{\sum_{i=1}^{n}(X_i - \overline{X})(Y_i - \overline{Y})}{\sqrt{\sum_{i=1}^{n}(X_i - \overline{X})^2 \sum_{i=1}^{n}(Y_i - \overline{Y})^2}}$$

A raw data "calculating" formula for r is

$$r = \frac{n\sum_{i=1}^{n}X_iY_i - (\sum_{i=1}^{n}X_i)(\sum_{i=1}^{n}Y_i)}{\sqrt{[n\sum_{i=1}^{n}X_i^2 - (\sum_{i=1}^{n}X_i)^2][n\sum_{i=1}^{n}Y_i^2 - (\sum_{i=1}^{n}Y_i)^2]}}$$

Do you think you could find the "deviation score" formula for r using the first equation for r above?

Let's see. Deviation scores are $x_i = X_i - \overline{X}$ and $y_i = Y_i - \overline{Y}$, so if I substitute x_i and y_i for $X_i - \overline{X}$ and $Y_i - \overline{Y}$ respectively, that should do it.

Correct. When this is done the formula using deviation scores is simply

$$r = \frac{\sum_{i=1}^{n} x_i y_i}{\sqrt{\sum_{i=1}^{n} x_i^2 \sum_{i=1}^{n} y_i^2}}$$

These are awful-looking expressions. Is there some pictorial or graphical way to compare relative degrees or magnitudes of r?

Yes, and the technique is called "scatterplots" or "scatter diagrams." It consists of plotting the pairs of X, Y data points on a rectangular coordinate system such that each point is represented by a pair of data values. Examples of possible scatter plots are given in Figure 6.

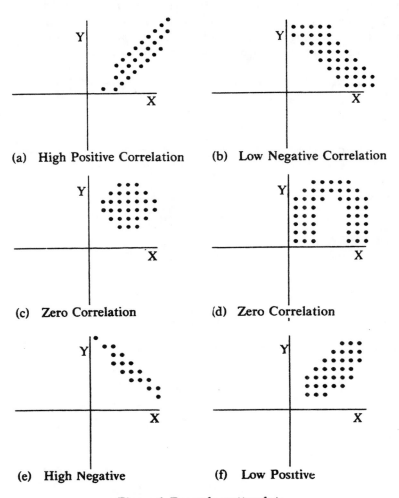

(a) High Positive Correlation   (b) Low Negative Correlation

(c) Zero Correlation   (d) Zero Correlation

(e) High Negative   (f) Low Positive

Figure 6. Example scatterplots.

You will notice that both (c) and (d) result in near zero correlations, since neither indicates a *linear* relationship. The scatterplot (d) indicates a quadratic relationship but since PPM deals only with *linear* relationships, its r value is still zero.

May I have a numerical example with the PPM?

Certainly. Consider the situation where you had attitude (X) and performance (Y) scores and wished to find the degree of linear relationship between them. Let the following set of eight pairs of scores represent the test scores of eight students on these two variables. Calculate r for this data.

Student	Attitude (X)	Performance (Y)
a	10	7
b	9	12
c	7	11
d	6	6
e	5	2
f	2	1
g	1	4
h	0	5

Answer: r = .67

It is apparent from the formulas and the pictures that a high PPM correlation (negative or positive) indicates that the points bunch-up close to a straight line. But of what value is this to me?

A high positive correlation means that the X and Y scores generally line up in order together. That is, large values on X are generally associated with large values on Y and likewise for small values of X and Y. A high negative correlation means that high values on one variable are generally associated with low values on the other variable.

For either high positive or high negative correlation this means that scores on one variable *could* be predicted with a good deal of accuracy from knowing scores on the other variable. A low correlation simply says that the predictions would most likely be poor ones in general.

O.K., prediction is something I would like to do. How do I go about making the prediction of one score from another?

This involves the use of a technique called *regression analysis* and requires the formulation of a "best" linear equation (regression equation), so that predictions can be made from it.

You mean regression analysis is simply finding an equation for predicting?

Yes, and since you want to make accurate predictions I'm sure that you would like to know that you are using the best prediction equation available. Right?

Right.

This being the case, we need to define "best." "Best" here will mean "least squares best" and will result in a line which is closer to more observed points than any other line.

This seems like an O.K. line, but how do I find the line? There are so many lines which could represent the data points.

A picture will help to clear up the technique of least squares which will produce the best straight line. Assume that you wish to predict Y from X. The line equation for this is called the *Y on X* line and will, in general, be different, as you will see, from the X on Y line. For the sake of simplicity assume there are only eight pairs of data points and their scatter plot is as shown on page 26.

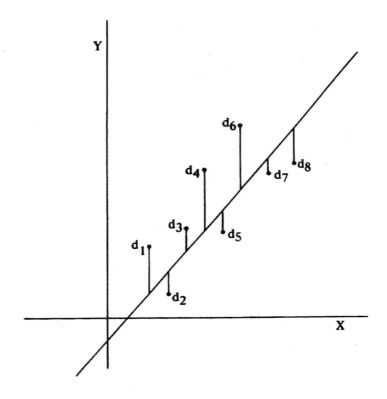

The $d_1, d_2, \ldots, d_8$ represent the vertical distances from each data point to any line through the scatter plot. The least squares line, denoted $\hat{Y} = a + bX$, will be the line such that the sum of squared distances to it will be less than the sum of squared distances to any other line. Notationally, the least squares technique will find the line, $\hat{Y} = a + bX$, such that

$$\sum_{i=1}^{8} d_i^2$$ is the smallest possible value. **(The $\hat{Y}$ denotes a predicted value)**

Now it is apparent that such a line will be a pretty good representative line.

Agreed, but I must repeat, how do I find it?

Well, it can be assumed that you are not interested in the technical details since differential calculus is used to find the values "a" and "b" in the line equation. But you would be satisfied to know that mathematics can produce the values "a" and "b," thereby giving you the equation of the best straight line.

I would be satisfied, so give them to me, already!

The formulas for a and b are

$$b = r \frac{S_Y}{S_X}$$

$$a = \overline{Y} - r \frac{S_Y}{S_X} \overline{X},$$

where $S_Y, S_X, \overline{Y}, \overline{X}$ are the sample standard deviations and means of the two variables and r is as previously defined. Thus, in order to find the best straight line to predict Y from X you need only solve for a and b above and substitute in the equation

$$\hat{Y} = a + bX.$$

I'm ready. Give me some numbers.

O.K., use the data and the correlation results on page 24 and find the regression equation to predict performance from attitude. Use the equation to predict a performance score for a student whose attitude score is 8.

Answer: $\hat{Y} = 2.5 + .70X$; for $X = 8$, $\hat{Y} = 8.1$

MEASURES OF PREDICTABILITY

You said the line for predicting Y from X and the line to predict X from Y were not the same generally. How come?

The answer is quite straightforward. To find the Y on X line you found the line which produced the minimum of squared *vertical* distance to the line. To find the X on Y line you need to find the line which produces the minimum sum of squared *horizontal* distances. These two lines will be the same *only* if $r = \pm 1$. (You have to take this on faith at present. In fact, the angle between two regression lines is a measure of the correlation. The smaller the angle, the larger the correlation and vice versa.)

Couldn't I just relabel the variables so that I would not have to worry about but one regression line at a time?

Sure, since you generally want to predict only one particular variable (*dependent variable*) from the other variable (*independent variable*).

Previously, we discussed high correlation implying good predictions and low correlation implying poor predictions. After I get a regression line to predict Y from X, how do I measure whether the predictions were "good" or "close" to the original data?

Good question. The answer is that there is another statistic to give an indication of how good your predictions were, using the regression equation, i.e., a *measure of predictability*.

The statistic used is called the Standard Error of Estimate and is denoted

$$\text{S.E.E.} = S_Y\sqrt{1 - r^2}.$$

It is "sort of" the average distance your observed scores are from your predicted scores.

Problem: Find the S.E.E. for your equation to predict performance from attitude, using the previous data.

Answer: 2.93

I now have learned how to calculate r, the regression equation and S.E.E. What does all this mean?

Basically, r indicates whether your predictions using the linear regression equation *will be* good ones or not; the regression equation gives the technique for *making* the predictions, and S.E.E. tells you how good your predictions *were*. In absolute terms they mean no more in value than $\overline{X}$ or S. Only in relative terms will they be interpretable. For example, an r of .72 is neither "big" nor "little" but, relative to another correlation on the same variables, it could be larger or smaller. Likewise an S.E.E. of 2.93 is, by itself, neither large nor small, but it could be relatively small if all previous studies had produced values over 5, for example.

The values r and S.E.E. are purely descriptive, like $\overline{X}$ and S, and should be used as such. Inferences relative to their population values can be made and will be done with r, later on.

If I found a PPM coefficient of .80, could I say that this indicates a linear relationship twice as large as a coefficient of .40?

NO.

Could I say that, in predicting one variable from another, I would be correct 80% of the time with an r value of .80?

You could say that—but you'd be incorrect.

How could I interpret an r value of .80?

It is known from statistical theory that an r value of .80 implies that approximately $(.80)^2$ (100) percent (64%) of the variance of one variable is predictable (accounted for) from the variance of the other variable. In other words, with an r value of .80 we know 64% of what we need to know to make a perfect prediction of either variable from the other.

If I get a high correlation between X and Y, does this say anything about X *causing* Y or Y *causing* X?

Absolutely not. Statistical correlation never implies causality; only degree of linear fit of the data points. Statistical indices *never* contain causality implications and *never* attempt to answer the question, "why?". Only the researcher can do this based on the information at hand.

You mentioned another measure of association called "Spearman Rho." What is it?

It is an index of association which makes no assumptions of linearity, or interval scale. It is appropriate to use if, for some reason, you are uncomfortable with the above assumptions.

Where does it come from?

The index, denoted r_s, is derived from the r formula by using *ranks* of the scores instead of the scores in the PPM formula.

What do you mean, "ranks of the scores"?

An example should make this clear. Consider the raw data on page 24. If for the X and Y variables separately we assign a "1" to the smallest score, "2" to the next largest, etc., we will have assigned ranks to the scores. The X and Y ranked scores will be as follows.

Student	Rank Attitude (X)	Rank Performance (Y)
a	8	6
b	7	8
c	6	7
d	5	5
e	4	2
f	3	1
g	2	3
h	1	4

Now that we have the scores ranked, do we just substitute these in the formula for r to find r_s?

You could do that but the arithmetic would be tedious. Besides someone named Prof. Spearman has already done that and derived a simple equation to short-cut all the work.

O.K., I'm glad he did it, but what's the simple formula?

The equation for r_s is

$$r_s = 1 - \frac{6 \sum_{i=1}^{n} d_i^2}{n^3 - n},$$

where the d_i are the differences between paired ranks, ignoring the sign. Using the data above, the unsigned differences are 2, 1, 1, 0, 2, 2, 1, 3. Do you see where the differences came from?

Yes, you simply found $8-6, 8-7, 7-6$, etc., for all eight pairs.

Exactly, and when these differences are squared and added together they become

$$\sum_{i=1}^{8} d_i^2 = 4 + 1 + 1 + 0 + 4 + 4 + 1 + 9 = 24$$

Thus, r_s becomes

$$r_s = 1 - \frac{6(24)}{8^3 - 8}$$

$$= 1 - \frac{18}{63}$$

$$= 1 - \frac{2}{7} = .71.$$

The answer for r_s is different from r as found on page 24. Is this expected or unusual?

Expected. Recall that we changed the scores to ranks, thus r_s could be quite different from or similar to r.

Can I say the same things about predictability with r_s that we said about r?

No, you cannot. There is no implied linear equation concerning which you could make a prediction.

Then what can I say about a numerical value of r_s?

Very little, except that it is a numerical index of the degree of rank association between X and Y and indicates how the scores go up and down together or in opposite directions. Some inferences can be made about r_s as in the case of r, but we'll not do that here.

Is the range of r_s from −1 to +1?

Yes, with negative values implying a reverse association between X and Y.

Is any causality implied with r_s?

No. It's the same as r in this regard.

Could I substitute it in the equation for S.E.E. in place of r?

The answer to, "could you" is yes. The answer to, "should you" is *no*. It is only an approximation of r_s if used in this manner. It may be a very poor or a good estimate and should not be used as a substitute for r.

It certainly is easier to calculate.

For a fact it is and should be used if you feel that any of the assumptions for r, previously mentioned, could not be met and did not wish to make linear predictions.

Just what are the assumptions for using r_s?

Both X and Y must be measured in at least an ordinal scale (so that the rank ordering makes sense).

GENERAL INFERENCE CONCEPTS

You have alluded to inferences several times. What kinds of inferences can I make from a sample?

Many, depending on the assumptions which are made concerning the data. (These assumptions will be discussed later.)

Could I have a few examples of what you mean by inference?

Yes, some examples of common *statistical* inference are:
 (a) The sample was drawn from a population with a mean unequal to 50;
 (b) Population means (variances, medians) are not equal;
 (c) Two variables are not independent;
 (d) The two samples were drawn from populations with unequal proportions of people with I.Q.'s below 80;
 (e) The sample was drawn from a population where the proportion of men is not equal to 1/2;
 (f) The correlation between achievement and verbal ability is larger than zero;
 (g) Boys have a higher correlation between achievement and verbal ability than do girls.

In addition to these inferential statement types, samples can be used to estimate parameters (e.g., population mean) and differences between parameters by a technique called *confidence intervals*. The confidence interval techniques are directly related to the techniques involved in making the above inferences.

These inferential statements are quite specific. What if I am not interested in these kinds of inferences?

If that is the case, then you probably should not be concerned with classical inferential statistics. If you do use statistics, however, then the types of statistical inferences which you can make are of this form, although the examples given are by no means exhaustive.

You keep saying, "statistical inference." Is there any other kind?

Yes, and it consists of inferential statements made as a *result* of statistical inferences. For example, if you statistically infer that the mean score of the population of students taught by method A is larger than the mean score of those taught by method B, you might infer that this is because the methods are different. "Methods are different," is not a statistical inference but an *experimenter inference* and may or may not be a valid inference to make, since there is more to "method difference" than arithmetic means of their respective population scores.

How do I proceed to assure myself that the samples taken allow me to make valid experimenter inferences?

One can never be certain that samples are of this type, but there are practical as well as statistical techniques which minimize the effects of uncontrolled factors. The most common, and probably the most effective practical and statistical method, is by using a *simple random sample* from a clearly defined and specified population. A simple random sample (or simply [no pun], random sample) of size n is a sample so selected that every possible sample of size n has an equal chance of being selected. Note that assuring that each element in the sample had an equal chance of being selected is a *consequence* of a random sample, not the *definition* of a random sample.

How do I get a random sample?

The easiest method is to assign every element in the population of interest a number and, using a random number table, select the number of observations desired.

How do I go about "clearly defining and specifying" the population?

This is done by listing all the characteristics which your population has, with the exception of the methods or treatments to be applied. For example, suppose two teaching methods are to be used on college freshmen. Now every conceivable characteristic which describes the student population should be specified, like age, sex, achievement level, etc. After the characteristics have been described, then random samples of freshmen are drawn (preferably equal in size) from the population of freshmen. One group is taught using method A and the other group method B. Any inferences thus made, are limited to the previously defined population of freshmen with some assurance that extraneous factors affecting the methods have been offset by the random nature of the samples. (See Cochran [1963] for additional help on special types of sampling techniques other than simple random sampling.)

How large a sample should I obtain?

The stock (and often unsatisfactory) answer usually given to this question is, "as large as possible." The sample size does, however, depend on what statistical inferences you wish to make and the statistical techniques used to make them. Given these specifications, there are several parameter values which must be estimated (or approximated) before one is sure of an adequate sample. A detailed answer to this question must wait until specific inferences and techniques (called *inferential tests*) are outlined and examples given.

You stated that the kinds of inference made depended on assumptions relative to the data. Do you mean that the data assumptions dictate the inference made?

Yes, the kinds of inference you can make using parametric statistics, for example, are quite different from inferences using some nonparametric statistics since they have different underlying assumptions.

What's the difference between parametric and nonparametric statistics?

They are two sometimes quite distinctive approaches to describing and making inferences from sample data. Parametric statistics generally involve a more stringent set of assumptions and generally have inferential statements which contain parameters; i.e., quantitative population values (Statement [a] on page 34 is a parametric example). The less restrictive set of assumptions allows the nonparametric inferences to be more generalizable. This may, or may not, be desirable, depending on the specificity needed by the researcher in his inferences. The exact distributional form (shape) of the population must be known in parametric testing, but not in nonparametric testing; thus the name "distribution-free" is often used in addition to "nonparametric." (Statement [c] on page 34 is a nonparametric inferential example.)

Is one type of statistical inference better than another?

There is no such thing as "better" in deciding between parametric and nonparametric inferential tests. Each has its own set of assumptions and each is valid for its own assumptions.

How does one decide which to use—parametric or nonparametric?

A serious researcher does not choose one or the other but entertains *both* types of inferential tests plus anything else that makes sense. Since one is not "better" than another, the selection of one must be made on the basis of, "which set of assumptions is more *defensible* from a research viewpoint"? For example, if you were hardpressed to believe that (1) the populations were normal, (2) variances were equal and (3) the data were interval in scale, then to make a test which assumed *all* these properties is not as defensible as conducting an analogous test which made *none* of these assumptions. (The word "normal" above refers to a specific, symmetrical bell-shaped distributional curve with which you may be familiar. It has particular and peculiar mathematical properties which make it a desirable distributional form to assume in classical parametric statistical work. It will not be discussed here, only assumed. Any basic statistical text will provide an adequate familiarization if needed, or see Appendix D.)

You have referred to "inferential tests" as "techniques." Are they techniques for making statistical inferences?

Exactly, and the technique involves the use of statements called hypotheses. The word "test" here is the process whereby the researcher rejects an hypothesis or not.

You mean a statistical inferential test result is a simple statement of "reject" or "not reject"?

Yes, and this should make it clear why a concise, testable statement of the hypothesis must be made and understood. For example, the inferential statement (a) on page 34 is the result of testing and rejecting the hypothesis, "The sample was drawn from a population with a mean of 50." A test result of "reject" is called a statistically "significant" result.

Do all those inferences on page 34 have similar testable hypothesis forms?

Yes, and the standard form for them is called the "null" hypothesis, denoted H_o, which generally is a statement of no difference, equal distribution, no association, no correlation, etc. The exact form of the hypothesis to be tested will depend on the assumptions relative to the testing technique used and may result in testing several analogous forms of H_o.

What is an "analogous H_o"?

An example is in order. Suppose you are testing:

H_o: $\mu_1 = \mu_2$ (i.e., two population means are equal)

and have met the assumptions presented on page 36. If you cannot defend these assumptions, then an analogous nonparametric H_o would be:

H_o: Samples are drawn from the same population.

The latter hypothesis could be tested with one of several nonparametric tests. Note that if the parametric assumptions are met, then the two H_o's are equivalent, since $\mu_1 = \mu_2$ and $\sigma_1^2 = \sigma_2^2$ forces the two normal populations to be identical. If the parametric assumptions are not met then the two hypothesis sets are not equivalent.

What if some assumptions are met and some are not?

In this case, both techniques (parametric and nonparametric) could be applied with a report of the assumptions of each test given. If the reader, upon seeing both results, wants to argue robustness of parametric tests and take that result, or argue for generality and take the nonparametric result, he may do so.

What is "robustness"?

Robustness is the degree to which a test is still appropriate to apply (with certain restrictions) when some of its assumptions do not hold. Some tests are more robust than others (for the same violation of assumptions) and some are robust with failures of single assumptions but nonrobust when combinations of assumptions are not met (see Glass, et al, 1972, for further discussion of robustness).

It is not clear to me how I would proceed to obtain testable hypotheses from a situation. May I describe a situation to see what kinds of hypotheses it yields?

Sure, go ahead.

I am interested in assessing the attitude of students toward a basic statistics course. It is of some interest to assess the attitude of those who pass and those who fail the course, before and after the course. Also I would like to say something about the relation between post-test performance and attitude. What are some possible testable hypotheses relative to this situation?

Before listing some testable hypotheses there are some *general research* points which have to be considered prior to conducting and interpreting any statistical test. Here are a few:

(a) You must have an operational definition of attitude and performance.

(b) You must have valid and reliable instruments to measure attitude and performance.

(c) You must clearly specify the who, when, where and how of the conduct of the statistics course.

(d) You must have some idea of the physical difficulty involved in the collection of the data.

After giving considerable thought to the above points, I feel I can assume that:

(a) Attitude and performance will be defined by scores on their respective measuring instruments. Scores will range from 0 to 100 on both instruments.

(b) All instruments are valid, reliable and simple to administer.

(c) Random sample data will be collected on all the students in the current academic year only.

(d) Instructor and meeting-time variables will be eliminated since it is an individualized course with no instructors.

Now what?

O.K., let's list some possible testable null hypotheses which could come from this situation. You will notice the similarity between some of these "null" hypotheses and their counterparts, the statistical inferences, denoted H_a (*alternate hypotheses*), on page 34.

(a) H_o: The sample of performance scores was drawn from a normal population with a mean of 80. (Note that 80 is an arbitrary number here and would have to be set by the researcher for some reason or other.) The symbolic form of this null is

H_o: $\mu = 80$.

(b) H_o: The sample of performance scores was drawn from a normal population with a variance of 100. (Here 100 is arbitrary, as 80 was above.) The symbolic form is:

H_o: $\sigma^2 = 100$.

(c) H_o: The population proportion of students with "bad" attitude after the course is equal to 1/2. Symbolically,

H_o: $P = 1/2$.

(d) H_o: The population variance for performance is the same for those with bad attitudes as for those with "good" attitudes. Symbolically,

H_o: $\sigma^2_{bad} = \sigma^2_{good}$

(e) H_o: The population mean performance is the same for those with bad attitudes as for those with good attitudes, or,

H_o: $\mu_{bad} = \mu_{good}$

(f) H_o: The population proportion of students with good attitudes is the same among "high" performers as among "low" performers, or,

H_o: $P_{high} = P_{low}$

(g) H_o: Performance and attitude are independent.

(h) H_o: Males have the same correlation between performance and attitude as do females. Symbolically,

H_o: $\rho_{males} = \rho_{females}$

Are these *all* the inferential hypotheses for the situation I outlined?

No, they are only a sample of possible ones. You also need to know that there may be several ways to test some of the above, depending on additional assumptions which need to be made.

Suppose only one or two of the hypotheses are of interest to me. Do I have to test them all?

No, you should have decided which ones you wanted to test *before* your study was designed, then built the investigation around what you wanted to test.

Suppose what I want to know cannot be tested by this limited number of hypotheses?

That can very likely happen, but you should not let the statistical technique dictate what you investigate. In other words, do not test an hypothesis simply because a statistical technique is available for that hypothesis. As a researcher you must decide what you want to do and, if there are statistical techniques available, use them. From this point on, it will be assumed that what you want to test *can* be tested with standard statistical techniques even though this may not always be the case.

Assuming that what I want to test can be tested with standard techniques, how do I proceed to test an hypothesis?

There are some *general* concepts and some *hypothesis-specific* concepts. The former are applicable to all standard statistical tests and the latter vary from hypothesis to hypothesis. The general concepts will be given here and the hypothesis-specific concepts will be utilized as each hypothesis and the test is entertained.

General concepts for hypothesis testing.

In conducting classical hypothesis tests a researcher begins by stating the null hypothesis, H_o, and the alternate hypothesis, H_a. The null has been previously discussed and the alternate is generally in the form of the statistical inference you wish to make. For example, one could state

H_o: $\mu = 80$
H_a: $\mu \neq 80$

where μ is the population average score on a performance test and 80 is arbitrarily set by the researcher. Here the researcher is saying that he believes μ is not equal to 80 but has no idea (or does not care) whether μ is greater than or less than 80, only that it is different from 80.

Note: The alternate, H_a, could be stated $\mu > 80$ or $\mu < 80$, depending on the inference desired, but should not be so stated unless the researcher is relatively certain of the direction the data will take or has interest only in that specified direction. A test with an alternate of $\mu > 80$ or $\mu < 80$ is called a *one-tailed test* (directional) because the direction of the result is predicted, and a test with an alternate of $\mu \neq 80$ is called a *two-tailed test* (nondirectional) because the direction is not predicted.

This seems straightforward enough. What do I need to know now that I have my H_o and H_a?

There are four basic entities or values which are inextricably related and which must be seriously considered if one is to conduct, interpret and report any statistical hypothesis test, either parametric or nonparametric. They are *alpha* (α), *power, effect size* and *sample size*. Their definitions, via the example hypothesis above, will be discussed here and their general interrelationships will be discussed later. Two of the values, α and power, relate to "actions" or decisions taken as a result of any statistical test and the other two, ES and n, directly affect these actions and thereby become crucial.

1. Alpha (α)

This value is the probability of a Type I error, where a Type I error is defined as the action of rejecting H_o when H_o is true. For the example hypothesis above, Type I error would be the error of saying, "μ is not equal to 80," when, in fact, $\mu = 80$. The relative frequency with which the researcher wishes to make this type of error is α. In long range terms it is the proportion of times (in tests exactly like the example hypothesis and using the same sample size) the researcher will commit this error. Tradition has almost decreed that it be set at .05 or .01 but the setting of α is purely arbitrary.

2. Power.

Power is the probability of the action to reject H_o when, in fact, H_o is false. Power is *not* the probability of an error but the probability of a correct action. In the example hypothesis, this correct action consists of saying, "μ is not equal to 80," when μ is not equal to 80, which is exactly what the researcher wants to do. This makes it understandable why the word "power" is used for the probability of this action. Like α, power is set arbitrarily by the researcher but, unlike α, has no traditional crutches on which to lean. By its very definition, it should be set as close to 1 as possible since there is no debate that "rejecting H_o when H_o is false," is the action desired in general by researchers or else they would not be testing hypotheses.

Power and α may be set by the researcher totally independently of each other but this may not be the most desirable method as we will see when the interrelationships are discussed. They must, however, be set prior to any data collection with power being set in conjunction with a specific effect size.

3. Effect size.

Effect size (denoted ES) is, according to Davies (1961), Cohen (1969) and other writers, the size of the treatment effect the researcher wishes to detect with probability power. It is obvious that when H_o is false, H_a is true and when

H_a is true μ and 80 are different. The researcher's judgment as to how large this difference is, or how large it must be before the researcher calls it important (nontrivial) enough to be detected with probability power, is ES. For example, in the performance score hypothesis, suppose a researcher believed that when μ and 80 were 5 units apart this represented a "real" difference and anything smaller than 5 was trivial. Further assume that the researcher wished to detect such a difference (or larger) with probability power = .90. That is, at least 90% of the time the researcher wants to reject H_0 (accept H_a) when μ and 80 are further apart than 5 units. Here, ES is 5.

Effect size, like α and power, is also arbitrarily set *a priori* by the researcher but involves a consideration of what the researcher wishes to find in the study as well as what is, to the researcher, important and what is trivial. It is hard to believe that these considerations are not a regular ingredient of all planned, serious research. Effect size is only a quantification of these considerations.

Cohen (1969) has simplified the quantification of ES by proposing that ES be expressed as a function of the overall population standard deviation, σ, whenever feasible. He has further proposed "small," "medium" and "large" ES values for most standard statistical tests relative to psychological research and Tversky and Kahnman (1971) have termed them "plausible" values. For the example hypothesis given above, Cohen (1969) suggested $.25\bar{\sigma}$, $.50\sigma$ and $.80\sigma$ respectively for small, medium and large ES values. The reader will note that this procedure makes the ES metric-free (independent of the measurement scale used) as well as eliminates the need to know or estimate the true population standard deviation.

You should not confuse effect size (that which is important to detect) with "significant" as used concerning a test result. Statistical significance; i.e., rejecting H_0, has nothing to do with ES since the former is an action taken as a result of the data and the latter is a judgment as to what is important in the view of the researcher. Some researchers have behaved as if they believed that statistical significance meant that the data was important but you should be aware of the fact that one could have statistical significance of trivial treatment effects. For example, an observed difference of 3 I.Q. points could be statistically significant but if you, the researcher, had previously stated an ES of 5 I.Q. points (in order to estimate your sample size) this observed difference of 3 may be trivial even though statistically "significant."

4. Sample size.

Sample size, denoted n, is the last but not the least of the values to be considered prior to any data collection. A typical first question after establishing an H_0 is, "How large a sample do I need?" The customary, but unsatisfactory, response you will recall is, "as large as possible." This need not be the case, however, because there is a very definite procedure for finding n for almost any classical hypothesis test and it involves the three previously discussed values: alpha, power and effect size.

It is well known, from the applied and theoretical statistics literature, that n is a function of α, power, σ and ES and, for almost all statistical tests, can be expressed by fairly simple formulas. (See Dixon and Massey [1957], Davies [1961] and Cohen [1969] for example formulas for n.) The researcher need not be concerned with several different formulas for the different hypothesis tests since Cohen (1969) has tabled a large range of n, α, power and ES values for most standard statistical tests, along with presenting discussions and interpretations of the interrelationships between α, power, ES and n. (An example formula for sample size calculation is given in Appendix E for a specified hypothesis test to be discussed later.)

You discussed a Type I error above. Is there a Type II error also?

Yes, and it is defined as the error of failing to reject H_o when H_o is false. The probability of this error is β and $1 - \beta$ = power; since power, as you will recall, is the probability of rejecting H_o when H_o is false. Obviously, when β is small, power is large and vice versa, but *α and β may both be small,* or large, depending on the sample size and effect size used.

I assume these relationships between α, power, ES and n are quite mathematical. But what is the general nature of the relationships in terms I can understand?

Let us consider the previous situation involving a test of

H_o: $\mu = 80$
H_a: $\mu \neq 80$

Now, if it is desired to detect almost any difference between μ and 80, no matter how small, then the researcher must be willing to obtain large samples, since he is assuming that μ and 80 are close together and he will need lots of information to detect this difference. On the other hand, if only gross differences are to be detected, then a smaller number of observations will be sufficient to detect the difference desired.

The relation between ES and n can be understood by considering the analogy of a needle in a haystack vs. a basketball in a haystack. If you want to detect a needle you must be prepared to divide the haystack into lots of segments (large number of observations) but if you want to find a basketball, the haystack need not be divided into as many segments (smaller number of observations). The point is, a researcher must know or estimate what degree of effect (needle or basketball) he wants before he can know whether he needs 10 or 10,000 observations.

The sample size also depends on the α and power $(1 - \beta)$ levels. Obviously, if one desires to make only a very small proportion of the Type I and Type II Errors, then he must pay for this virtual "error-free" state by collecting large amounts of data. Bear in mind, the purpose was, and is, to make an inference about an infinite population of numbers and the way to minimize both kinds of error is to take large samples (all samples are assumed, of course, to be random).

Thus, it appears that n goes up as ES, α and β go down, thereby causing power and n to go up together since $(1 - \beta)$ = power. *Only* for n and ES fixed will α and β vary inversely. Otherwise, they can both be set small (desirable state) or both large (undesirable state).

Do these same general relationships hold for all statistical tests?

Yes, but the degree of relationship and relative size of ES will differ from test to test due to the different interpretations of ES for the different tests. For example, ES has a different meaning for the test

$H_o: \mu = 80$ than for the test $H_o: \mu_1 = \mu_2$

$H_a: \mu \neq 80$ $H_a: \mu_1 \neq \mu_2$

You were discussing general concepts for testing hypotheses and have mentioned stating the H_o and H_a, setting α, power and ES and finding the minimal sample size. What now?

In general, the statement of the H_o determines the type of test or tests to be used, so assume that α, power, ES and n have been decided on for the test of

$H_o: \mu = 80$
$H_a: \mu \neq 80$

Now, the basic idea underlying inferences via hypothesis testing is to let the data have a chance to reject (or not) H_o for you in keeping with the error rates, ES and n, which have been previously set. In order to do this we must find out how *likely* the data we collect is when H_o is true. Then, if it is highly unlikely that the data collected was from a population with $\mu = 80$, for example, if P (data given H_o true) $\leq \alpha$, we can choose to reject H_o. If the likelihood of the data is $> \alpha$, then we can choose not to reject and reach what is called a "no rejection." So, the main thrust of inference is to find the likelihood of the data given H_o is true so a statistical decision can be made. This likelihood, or probability, is denoted "p" in most behavioral literature.

How do I find the likelihood of my sample data under the assumption that H_o is true?

The answer to this question involves a discussion of what is called "probability and sampling distributions." But for the present, finding the likelihood involves knowledge of the *sampling distribution* of the test "statistic" (a mathematical function of $\overline{X}$ and S) and the calculation of the probability of the statistic when H_o is true. The form and details of the distribution of the test statistic will be presented when we consider the specific tests in greater detail.

In the answer to a previous question you said we can "choose" to reject H_o if the likelihood, p, was less than or equal to α. Does this mean we could choose to do otherwise?

Yes. Rejecting H_o when the likelihood is less than or equal to α is simply a statistical decision convention. Note also that a statistical decision (or inference) is not necessarily the same as an experimenter decision. A statistical rejection of $\mu = 80$ does not *guarantee* that $\mu \neq 80$, but only that if $\mu - 80 = ES$ we had a probability of at least $1 - \beta$ of detecting this or any larger differences. As a researcher, you may choose to do anything you want with the result of a test but it may or may not be statistically defensible.

SPECIFIC HYPOTHESIS TESTS

Would you conduct hypothesis tests for some of the H_o's previously cited so I can see how it's done?

Certainly, and they will be some of the most commonly occurring tests in applied statistics. This will by no means be all possible tests but will be sufficient for an introduction. It will be assumed for all the following tests that α, power, ES and n have been previously and satisfactorily determined and that all samples are random. The hypotheses for which tests will be conducted are:

a. H_o: μ = constant
b. H_o: $\mu_1 - \mu_2$ = constant
c. H_o: $\sigma_2^2 = \sigma_2^2$
d. H_o: $\rho = 0$
e. H_o: Two variables are independent.

Test statistic for H_o: μ = constant.

In this test the researcher is interested in testing that the mean of a population of scores, for example, on an achievement test, is equal to some constant, say 80. He does not believe this, but thinks, for some reason or other, that $\mu \neq 80$. There is a parametric test for this hypothesis called the *student's t-test* and is the most common of the parametric tests in behavioral research. The test statistic is

$$t = \frac{(\bar{X} - \mu)}{S/\sqrt{n}},$$

where $\bar{X}$ is the sample average and S is the standard deviation of the sample. (This t is the mathematical function of $\bar{X}$ and S mentioned previously.) The sample size, n, was determined, as you recall, by setting α, β and ES and referring to a source like Cohen (1969). The effect size (ES) for this test is the magnitude of the difference between μ and 80 you wish to detect with probability power.

It can be decided on, with some assistance from Cohen (1969), by asking yourself, "How far apart must μ and 80 be before I can call the difference nontrivial (important, meaningful, real, etc.)?" Cohen suggests $.25\sigma$, $.50\sigma$. and $.80\sigma$ for small, medium and large ES's respectively, but these are only suggested values.

Assumptions of the test for H_o: μ = constant.

1. The achievement scores are normally distributed with mean μ and variance σ^2.
2. The scale of measurement is at least interval.
3. The population variance σ^2 is unknown (as is usually the case) but estimated by S^2.
4. Scores are randomly sampled.
5. Scores are independent, i.e., no score can affect, relate to, or be associated with any other score.

Likelihood under H_o: μ = constant.

If H_o were true then the test statistic for H_o: μ = 80 becomes

$$t = \frac{(\bar{X} - 80)}{S/\sqrt{n}}$$

As previously discussed, if one can find the probability (likelihood), p, associated with this statistic (a function of the data) under H_o then one can decide to reject H_o if $p \leq \alpha$ or not reject H_o if $p > \alpha$.

What do you mean "probability *associated* with this statistic"?

"Associated" probability means the probability, denoted p, of obtaining a t-value equal to, or more extreme than, the one found from our data when μ = 80. To illustrate, suppose

$\bar{X} = 83$
$S = 8$
$n = 64$ (These are made-up numbers—you missed nothing.)

Now the calculated t-statistic is

$$t = \frac{(83 - 80)}{8/\sqrt{64}} = 3$$

You must now ask, "What is the probability, p, of obtaining a t-value more extreme than, or equal to 3 when μ = 80?" If $p < \alpha$ then you can reject H_o and conclude that $\mu \neq 80$. If $p > \alpha$ then you do not reject and conclude nothing (remember this was a "no decision"). Note that if $\overline{X}$ had been 77 you would have had to ask the same question concerning a t of -3.

Since it is known from statistical theory that the probability of obtaining a t-value more extreme than 3 (larger than 3) is the same as obtaining a t-value more extreme than -3 (smaller than -3) you need only know the probability for the positive case, i.e., you need the p for a t-value larger than, or equal to, 3.

How do I find the probability of a t-value larger than 3?

Actually, you don't. What you do, is find the t-value which is associated with a probability of α and ask "Is my calculated t larger or equal to this t associated with α, denoted t_α?" If it is, then reject H_0. If not, then do not reject H_0. (The extensiveness of the tables makes it easier to do it this way rather than finding p, but it is equivalent.)

Big deal. How do I find t_α?

That's the simple part. You consult a table of t_α values for your example. What's not so simple is where the table comes from. You may breathe easy, however, since you need not develop the table in this material, only consult it. You need only know that a t-distribution is very similar in appearance to a normal distribution and that the larger the obtained t-value, the smaller is its associated probability. This is why finding t_α and comparing your calculated t to it is equivalent to finding the probability of the calculated t and comparing it to α (see t-test Table A).

O.K., how do I use the table?

The table is constructed for both two-tailed and one-tailed tests. The column headed df denotes the "degrees of freedom," which for this test is $n - 1$. The other column headings denote the α levels selected *a priori* by you. You simply select the column for your α level and read down until you are opposite your degrees of freedom. (The table is not complete so you may have to approximate by using the next smaller df but this is O.K.) If your calculated t-value is greater than or equal to the table entry, then reject H_0, otherwise do not. Using the data on page 47 you see that $df = n - 1 = 63$. Suppose you had selected an α value of .05, then, since your calculated value of 3 exceeds the table value t_α of 2, you may reject H_0 and conclude that $\mu \neq 80$. If, however, you had previously selected an α of .001, then you would fail to reject H_0 since your calculated t value of 3 is not greater than the t_α of 3.460. (Note that the same conclusions would be reached if the calculated t had been -3 since you are conducting a two-tailed test and thus not interested in the direction of the difference between μ and 80.) Also note that a test with $\alpha = .05$ is a different test from one with $\alpha =$

.001 since different sample sizes would be necessary to maintain the same power and ES.

What's this degrees-of-freedom business?

That's a difficult question and cannot be answered fully in this text. Basically each t-distribution is different as different sample sizes are used. For example, if n = 26 is used, this generates a complete distribution of values for t (examples are found in the row across from df = 25). If n had been another value, say n = 16, then this generates another t distribution. The degrees of freedom tells you which distribution you are consulting in the table. Another way to remember how to find df is to ask, "How many observations are free to vary among the 64 if I know that $\overline{X}$ is fixed at 83?" Since, as you recall

$$\overline{X} = \sum_{i=1}^{64} X_i/64 = 83,$$

then 63 scores (n – 1) can be anything and the 64th one can be found from them using $\overline{X}$ = 83. Thus, degrees of freedom is n – 1. If your df is not shown in the table then use the next smaller df value in the table or interpolate.

O.K. now give me a problem and let's see if I can work it.

Alright, try this one. Suppose you want to test

H_o: $\mu = 32$
H_a: $\mu \neq 32$

and find that $S^2 = 36$ and $\overline{X} = 30$ when n = 100. Assume further that you wish to conduct the test with $\alpha = .01$. What is your statistical decision?

Answer: t = 3.33, df = 99 (use df = 60), thus your decision should be to reject H_o. Now do the next problem which involves a bit more computation and review of $\overline{X}$ and S^2.

The following is a set of different scores resulting from a pre-post testing situation. You would be interested in testing that the population mean of all these difference scores is zero, i.e., in testing

$H_o: \mu = 0$
$H_a: \mu \neq 0$

Assume the test is to be conducted with $\alpha = .02$. The difference scores are

2, −6, −3, 4, 2, 4, 1, 5, 0, −4.

What is your statistical decision? (Note that the sample size is too small to be very meaningful, but that this is only an example.)

Answer: t = .446, df = 9, thus do not reject H_o.

I feel I can do hypothesis tests of this type, but what if I wanted to make a one-tailed test–say,

$H_o: \mu = 0$
$H_a: \mu > 0?$

The procedure is exactly the same except for the reference in the table. In this case the t_α with which you compare your calculated t is found under the column headed "One-tailed." Note that a one-tailed $\alpha = .05$ test uses the same table values as a two-tailed $\alpha = .10$ test.

A word of caution: This one-tailed test should be conducted *only* if your interest is in the hypothesized direction exclusively. In the immediately preceding example this means $\overline{X}$ has to be > 0 or no rejection is possible. Common sense will tell you why this is so: There's no way to conclude that $\mu > 0$ if the estimate you have, i.e., $\overline{X}$, indicates otherwise.

Earlier you said that confidence intervals were directly related to tests of hypotheses. What are they?

The technique of confidence intervals is primarily thought of as a parameter estimation technique but it may be utilized in conjunction with, or instead of, hypothesis testing. They involve different sample size requirements, but these will not be discussed here. See Brewer, 1986 for details.

For example, suppose you wished to estimate μ with an interval instead of using only the *point estimate*, $\overline{X}$. The idea of confidence intervals is to construct an interval which consists of two numbers denoted (L_1, L_2), such that a high proportion of all such intervals would "surround" or contain the true value μ. The interval actually constructed is obviously only one of an infinite number of such intervals but one places confidence in it because of the high proportion of all intervals *like it* which would contain μ.

To illustrate, suppose you wanted a 95% confidence interval on μ. That is, you wanted to construct an interval (consisting of two numbers) such that 95% of all such constructed intervals (using the same sample size) would contain the population mean, μ. (Obviously, 5% will *not* contain μ). A 95% confidence interval (C.I.) for μ will be

$$\overline{X} \pm \frac{S}{\sqrt{n}} (t_{.05})$$

where $\overline{X}$ and S are as before and $t_{.05}$ is the table t-value for $\alpha = .05$ (since $1 - .95 = .05$). Using the data on page 47, the C.I. becomes

$$83 \pm \frac{8}{\sqrt{64}} (2) = 83 \pm 2$$

or the interval from 81 to 85, denoted (81, 85). There is no guarantee that μ is between 81 and 85 but 95% of all *such* intervals using sample a size of 64 will contain μ.

That seems like a reasonable way to estimate μ which utilizes the same information as in hypothesis testing. How are C.I.'s used in hypothesis-testing?

It's quite simple. If the hypothesized value of the parameter lies *outside* the constructed confidence interval then H_0 is rejected; otherwise not. In the example above, if H_0 had been $H_0: \mu = 83$ then you would not reject H_0 since 83 is not outside the interval (81, 85). If, however, you previously stated $H_0: \mu = 74$, then H_0 would be rejected since 74 is outside the interval (81, 85). Do not forget, however, that if hypothesis testing is to be conducted with C.I.'s then prior consideration must be given to α, β, ES and n for the particular two-tailed hypotheses to be tested as well as to the determiners of adequate sample size for confidence intervals.

Problem: Construct a 99% confidence interval on μ for the data on page 49.

Answer: (28.40, 31.60)

Can confidence intervals be constructed on other parameters like $\mu_1 - \mu_2$, σ^2, etc.?

Yes, and the same general procedure is followed using the information which would be necessary if you wished to hypothesis test for these parameters. We will not construct intervals for these other parameters but you should know that it is commonly done and that almost any complete statistics text will give you the proper formulas. The concept is the important thing here.

This hypothesis testing for one mean seems simple enough but suppose I wanted to compare two population means. Isn't this more complicated?

Not at all.

There are, however, two basic kinds of population means tests: Those which are made using *independent* samples and those made using *related* samples.

Independent samples are samples so selected that no observation in one sample can affect, relate to, or be associated with any observation in the other sample. An example of independent samples would be for one sample to be a random collection of I.Q. scores from one school district and the other sample randomly selected from I.Q. scores in another school district. In this case there would be no reason to believe the scores are related, so we could assume independence.

Related samples are samples which are not independent, i.e., they are samples so selected that you have reason to believe that the scores in the two samples are associated. For example, if you conducted pre- and post-tests on the same group of people then you would have reason to believe that the scores are related since they are on the same set of people. You could also obtain some degree of relatedness by pairwise matching two sets of people on all relevant variables. This latter approach may, however, be difficult since obtaining a match for any pair on all "relevant" variables might be next to impossible as the list of relevant variables could be quite long.

Whether the two samples are independent or related it is necessary always to have *intra-sample independence*, that is, independence between all observations within each sample. As a researcher you must be sure that each observation within a sample cannot relate to, be associated with, or affect any other observation in that sample. As you can imagine, it would be impossible to statistically analyze and interpret a situation, for example, where you knew half the students obtained their performance scores by cheating from the other half, i.e., scores are a function of other scores.

How would the test be conducted for two independent samples?

The test of

$H_o: \mu_1 = \mu_2$
$H_a: \mu_1 \neq \mu_2$

for two independent samples involves exactly the same kinds of data, the same test statistic and the same table values. Only the degrees-of-freedom values and the calculating formula will differ, thus affecting the place in the table you look.

Test statistic for H_o: $\mu_1 = \mu_2$.

As previously mentioned, the researcher here is interested in testing the equality of two population means using two independent samples. He also could test that the two means differed by some amount by testing H_o: $\mu_1 - \mu_2 =$ constant, but the procedure is the same. The test for equality is the most common and will be used to illustrate the technique.

Effect size for this test now is the answer to, "How far apart must μ_1 and μ_2 be before I can call them nontrivial?" (Or the size of the difference I wish to detect with probability power.) Cohen (1969) suggests $.25\sigma$, $.50\sigma$ and $.80\sigma$ respectively for small, medium and large ES. (You will see why there is only *one* σ when the assumptions are given below.)

The test statistic here is

$$t = \frac{(\bar{X}_1 - \bar{X}_2) - (\mu_1 - \mu_2)}{\sqrt{\left[\frac{(n_1 - 1)S_1^2 + (n_2 - 1)S_2^2}{n_1 + n_2 - 2}\right]\left(\frac{1}{n_1} + \frac{1}{n_2}\right)}}$$

where S_1^2 and S_2^2 are the two sample variances, $\bar{X}_1$ and $\bar{X}_2$ are the two sample means and n_1 and n_2 are the two sample sizes. It looks messy but you know how to calculate everything in it. It's only a matter of arithmetic.

Assumptions for the test of H_o: $\mu_1 = \mu_2$.

(1) The two populations of scores are normally and independently distributed with means μ_1 and μ_2 and variances σ_1^2 and σ_2^2.
(2) $\sigma_1^2 = \sigma_2^2 = \sigma^2$ but is unknown (now you see why there is only one σ).
(3) The scale of the measurement is at least interval.
(4) Scores are random samples with intra-sample independence.

Likelihood under H_o: $\mu_1 = \mu_2$.

If H_o is assumed to be true then the test statistic becomes

$$t = \frac{[\bar{X}_1 - \bar{X}_2] - 0}{\sqrt{\left[\frac{(n_1 - 1)S_1^2 + (n_2 - 1)S_2^2}{n_1 + n_2 - 2}\right]\left(\frac{1}{n_1} + \frac{1}{n_2}\right)}}$$

(Note that if the null hypothesis had been $H_o: \mu_1 - \mu_2 = 5$, then 5 would be subtracted from the numerator of the statistic above.)

From this point on the procedure is exactly as in the single sample test of $H_o: \mu = $ constant. One merely finds $\bar{X}_1, \bar{X}_2, S_1^2, S_2^2, n_1, n_2$ and plugs them into the above expression. This will give the calculated t value.

With what do I compare this calculated value since you said the degrees of freedom would be different?

Good question! The degrees of freedom for this test will be $n_1 + n_2 - 2$. Common sense again tells you this is reasonable since for the single sample it was $n - 1$, and here we have two samples, so $(n_1 - 1) + (n_2 - 1) = df = n_1 + n_2 - 2$. *Voila!* Knowing this and your preset α locates you exactly in the t-table. If your calculated t is larger than the table t_α, then reject. If not, then do not reject.

There doesn't seem to be too much to this test. Give me a problem and see how I do.

O.K., try this. Suppose $\bar{X}_1 = 100, \bar{X}_2 = 98, S_1 = 4, S_2 = 3, n_1 = 10$, and $n_2 = 17$. Assume that α was preset at .05. (Caution: These sample sizes may be too small to be very meaningful but are used since they simplify the calculations. See Appendix E for adequate sample sizes. Remember this is for illustration only.) What is your statistical decision for a two-tailed test?

Answer: $t = 1.48$, df = 25, thus your statistical decision should be not to reject H_o.

Now work a problem using raw data. Suppose two groups were given a simple motor task and their scores were as follows:

Group I	Group II
7	9
6	7
3	8
4	4
2	

Test: $H_o: \mu_1 = \mu_2$ $H_a: \mu_1 \neq \mu_2$

at $\alpha = .20$ and reach a statistical decision. (Here again these sample sizes may be too small to be meaningful but are examples only.)

Answer: $t = 1.83$, df = 7, thus reject H_o.

If I wanted a one-tailed test, is the table value for this two sample procedure the same as for a one sample test?

Yes.

All I need remember then, is that if the difference between $\bar{X}_1$ and $\bar{X}_2$ is not in the same direction as the alternate hypothesis, then I have no way to reject H_o?

Correct, but recall that H_a *must* be established *prior* to collecting any data.

What if I look at the data, then establish an H_o and an H_a to suit the data?

Then you're an academic crook! Anyone can fabricate a hypothesis set which will "reject" if the data is consulted first, but it will be meaningless and benefit no one.

The previous test was for independent samples. What if they are related samples, for example, pre- and post-test data on the same subjects?

In this case it is apparent that the difference scores (Post-Pre) constitute a single sample of scores, and you test this with $H_o: \mu_d = 0$, i.e., that the population mean of the score *differences* is zero. You have conducted a test of this type on page 50, and may not have known it. The reason this works is that $\mu_1 = \mu_2$ is the same statement as $\mu_1 - \mu_2 = 0$ and you recall that the mean of differences is the difference of the means, so, H_o is $\mu_d = 0$.

You mean that's all there is to testing for means in related or associated samples?

Basically, yes, and this should not surprise you since the differences are what you're interested in and they constitute a single sample.

What about testing that the variances are equal?

Test statistic for H_o: $\sigma_1^2 = \sigma_2^2$.

The test statistic for this hypothesis test is

$$F = \frac{S_1^2}{S_2^2},$$

where S_1^2 and S_2^2 are the sample variances of the two samples with S_1^2 denoting the larger of the two. (One of the sample variances obviously will be larger or no test would be conducted.) The effect size for this test is simply the size of the ratio σ_1/σ_2 which you wish to detect with probability power.

This certainly is a simple-looking test statistic. Do we calculate it and compare it with a table value as before?

Exactly. The only difference, however, for this test is that there are *two* degrees-of-freedom values. The degrees-of-freedom are denoted df_1 and df_2 to represent respectively the numerator and denominator degrees-of-freedom for the calculated F. It will be of no surprise to you now to know that $df_1 = n_1 - 1$ and $df_2 = n_2 - 1$. The F-test, Table B, pages 70–73, has the df_1 values as column headings and the df_2 values as row headings. Knowledge of df_1, df_2 and α will locate you the proper F_α with which to compare your calculated F value. If your calculated F is larger than the table F_α, then reject; otherwise do not reject.

Assumptions for the test of H_o: $\sigma_1^2 = \sigma_2^2$.

(1) The two populations, whence cometh the two samples, are normally and independently distributed with means μ_1 and μ_2 and variances σ_1^2 and σ_2^2.
(2) σ_1^2 and σ_2^2 are unknown and estimated by S_1^2 and S_2^2 respectively.
(3) The scale of measurement is at least interval.
(4) Scores are randomly selected with intra-sample independence.

By jove, I've got it! Let's do a problem.

O.K., Suppose you are testing

H_o: $\sigma_1^2 = \sigma_2^2$
H_a: $\sigma_1^2 \neq \sigma_2^2$

and found that $S_1^2 = 18$ and $S_2^2 = 9$ where $n_1 = 31$ and $n_2 = 25$. Also assume $\alpha = .10$. What is your statistical decision?

Answer: The calculated F value is F = 18/9 = 2. The table F_α for $n_1 = 31$ and $n_2 = 25$ is 1.94 ($df_1 = 30$ and $df_2 = 24$). Since your calculated value is larger than the table F_α, then reject H_o and conclude that $\sigma_1^2 \neq \sigma_2^2$.

The above test was two-tailed. What if I wanted to conduct a one-tailed test?

The procedure is the same as on previous tests, assuming the data is in the direction predicted by H_a. Note that there are only a few α values for the F-tables so in order to conduct tests with many different α levels, you will need to consult more extensive tables, which are available in the statistical literature.

In testing $\mu_1 = \mu_2$ previously, it was assumed that $\sigma_1^2 = \sigma_2^2$. Is this equality of variance test a test of this assumption?

Correct. However, one could conduct the test H_o: $\sigma_1^2 = \sigma_2^2$ with no thought for the t-test of H_o: $\mu_1 = \mu_2$. It would be a good idea, nevertheless, to precede every t-test of H_o: $\mu_1 = \mu_2$ with this equality-of-variance test (or commonly called the *homogeneity-of-variance* test) if normality can be assumed.

Example problem.

Conduct a homogeneity-of-variance test for the raw data problem on page 54 with $\alpha = .10$.

Answer: Do not reject H_o: $\sigma_1^2 = \sigma_2^2$, since F = 1.09, $df_1 = 3$, $df_2 = 4$

In the preceding tests the examples used had unequal sample sizes, yet you previously advised us to keep sample sizes equal if possible. Howcum?

In order to maximize the power of the test you should keep the sample sizes equal. For example, in a two-group t-test power will be greatest if the total sample is equally distributed (randomly, of course) to the two groups.

This is sometimes difficult to do since there are many possibilities for missing data especially when dealing with experiments in real life.

The examples previously provided were to give you practice with different degrees of freedom and the table values. The unequal samples as well as their size, were not necessarily realistic but were examples only to get the concept across. Did it work?

I think so.

Good. Now go on before you get stale.

Correlations are a big thing in my area. What about the test $H_o: \rho = 0$?

Test statistic for $H_o: \rho = 0$.

In this test the researcher wishes to test that the Pearson Product-Moment Correlation Coefficient, ρ, equals zero when he believes that the true correlation coefficient is not zero. For example, this test could be conducted if one believed that performance and attitude were linearly correlated and wished to support this statistically. The effect size here is simply the size of the population ρ you will allow to go undetected with probability β. Cohen (1969) suggests .10, .30, .50 for small, medium and large values. The statistic to be calculated is

$$t = r \sqrt{\frac{n-2}{1-r^2}},$$

where r is the sample correlation coefficient and n is number of pairs. Since this is the same t-distribution as before we need degrees-of-freedom, which for this test is $n - 2$. Once r is found, the procedure is the same as for any t-test, namely, calculate t and compare with t_α for df = $n - 2$ and α preset. That's all there is to it!

Assumptions for the test of $H_o: \rho = 0$.

(1) The two variables form a bivariate normal distribution.
(2) The measurement scale is at least interval.
(3) Pairs are randomly sampled with intra-sample independence.

Example:

Suppose you wish to test that the correlation between Math and English scores is zero using a sample of 51 students (i.e., n = 51). Make a statistical decision if you found the sample correlation coefficient, r, to be .60 and did not care about direction of the correlation. Let $\alpha = .01$.

Answer: Your test is

$H_o: \rho = 0$

$H_a: \rho \neq 0$.

The statistic under H_o becomes:

$$t = .60 \sqrt{\frac{51 - 2}{1 - (.60)^2}} = .60 \sqrt{\frac{49}{.64}} = 5.25$$

Now, consulting the t-table across df = 40, you see that 5.25 is larger than the t_α of 2.704. Therefore, reject H_o and conclude that $\rho \neq 0$.

Problem:

Test $H_o: \rho = 0$ using the following data with $\alpha = .10$, two tailed.

X	Y
1	2
2	4
3	3
5	1

Answer: Do not reject H_o since r = −.529 and t = −.88.

The form of hypothesis (e) on page 46 is quite different from the others. What's so special about it and why does it not contain parameters?

This hypothesis contains no parameters since it is a common example of a nonparametric test used with frequency data to test for independence between two variables. (Note that its parametric analog is the test $H_o: \rho = 0$, which you have previously conducted.) The data on each of the two variables is in the form of frequencies in discrete categories. For example, suppose one is interested in testing the independence between attitude and performance and that there were three categories of attitude: good, fair and poor; and three performance categories: high, medium and low. The null hypothesis H_o would be that attitude and performance are independent (unrelated, unassociated, etc.). Suppose further that 200 students were tested on performance and attitude with the following frequency outcomes.

Attitude

		Good	Fair	Poor	
Performance	high	25	13	6	44
	med	25	25	25	75
	low	10	21	50	81
		60	59	81	200 = Total frequency

The question which has to be asked now is, "What frequencies would one *expect* to see in a table like this if H_o were true?" That is, what would the frequency entries be in the long run, if performance and attitude were independent? If the *expected* frequencies were found to greatly different from the *observed* table entries, then we would suspect that H_o was false since the expected entries were the result of H_o being assumed true. A test statistic called the "Chi-Square" statistic is used to test if the differences between the *observed* and *expected* frequencies are large enough to warrant rejection of H_o.

Test statistic for H_o: *Attitude and performance are independent.*

The form of the test statistic is

$$\chi^2 = \Sigma \frac{(O_i - E_i)^2}{E_i}$$

where O_i and E_i are respectively the observed and expected frequencies in cell i and the sum is over all cells. (In our example there are nine cells.) As in all the previous tests you need only calculate χ^2 and compare it with a table value for some preset α. The Chi-Square Table C, gives the χ^2_α values for several α's and several degrees-of-freedom (df). For this test, the degrees-of-freedom are (categories of performance $-$ 1) (categories of attitude $-$ 1) = (3 $-$ 1) (3 $-$ 1) = 4.

Effect size estimation is not as straightforward as it is for the other tests but Cohen (1969) and Guenther (1977) offer some assistance.

Assumptions for Chi-Square independence test.
 (1) Each observation may be categorized into exactly one of the cells.
 (2) The classifications into the cells are independent.
 (3) Data is frequency in form.
 (4) Samples are randomly selected.

The test statistic seems O.K. and I have the O_i values, but how do I get the E_i values?

Let's take a particular cell and find what the expected frequency should be under the assumption that H_o is true. For example, take the cell "high performance, fair attitude." (You have observed 13 people to have these two characteristics.) How many *should* be seen in that cell if H_o is true? Well, if attitude and performance are independent, then having a fair attitude has nothing to do with having a high performance, so the probability of having a fair attitude *and* a high performance is simply the product of the probability of a fair attitude and the probability of a high performance. (This is a basic probability rule given in Appendix C.) Once we know the joint probability of, "having a fair attitude *and* high performance," to find the expected number (E) of people with these two characteristics we need only multiply this joint probability by the number of people available, which in the example is 200.

So the problem boils down to finding the two probabilities, namely the probability of a fair attitude and the probability of high performance. It is apparent that you do not know these probabilities exactly but estimates can be made from the observed data.

Common sense tells us that if 59 people have a fair attitude (independent of their performance) then 59/200 is a reasonable estimate of the probability of having a fair attitude. Likewise, 44/200 is a reasonable estimate of the probability of having a high performance (independent of attitude). The product (59/200) (44/200) is then the joint probability of being in this fair attitude, high performance cell. The expected number (E) of people in the cell is thus

$$(200)(59/200)(44/200) = \frac{(59)(44)}{200} = 12.98.$$

Recall from the data that the observed number of people in this cell was 13. For this cell, at least, what one should expect under H_o and what is observed are quite close.

The same procedure has to be used on all cells and you will save time by noting that in the above equation all one needs is the product of row and column totals for the cell of concern divided by the grand total. For another example, the cell, "low performance and poor attitude" has an expected frequency of

$$\frac{(81)(81)}{200} = 32.80.$$

The observed value was 50.

Problem: Finish finding all the expected frequencies in all the cells.

Answer:

Expected Frequencies

Attitude

		Good	Fair	Poor
	high	13.2	12.98	17.82
Performance	med	22.5	22.13	30.38
	low	24.3	23.89	32.80

A word of caution about the use of the χ^2 statistic with frequency data is in order. There should not be many cells with very small cell frequencies and *none* of the expected cell frequencies can be zero. The latter is obvious since E_i is the denominator of the test statistic.

Now that you have the observed and expected frequencies in all the cells, you need only calculate χ^2 and compare it with the table χ^2_α.

Problem: Find χ^2 and make a statistical decision if $\alpha = .05$.

Answer: Since $\chi^2 = 37.50$ and df = 4 reject H_0 and conclude attitude and performance are not independent.

We have conducted statistical tests for the hypotheses on page 46. Are there other statistical tests for these same hypotheses?

Yes, there are scores of other statistical tests for these example hypotheses. Some have stringent assumptions and some are relatively assumption-free.

Where would I find some of these tests?

Almost any textbook relative to *nonparametric* or *distribution-free* statistics would provide you with several alternative tests.

Do all the alternative tests involve the same basic philosophy, namely, setting up an H_0 calculating a statistic and rejecting or not?

Yes, but you must be aware that this inferential approach is based on only one philosophical viewpoint and that there are other philosophies which make equal sense to some researchers.

For instance?

An example of an inferential philosophy with a different viewpoint is the Bayesian philosophy. It is named after an English theologian who proved an elementary probability theorem on which the decision mechanism of the Bayesian philosophy rests. (The theorem is called Bayes' Theorem, naturally.)

BAYESIAN STATISTICS

What's Bayesian statistics all about?

There's no short answer to this question, but several pieces of literature are available if you are interested. Start with any *elementary* exposition of Bayesian statistics so that the mathematics of the technique does not cloud the concept.

Can you describe the basic difference between "classical" and Bayesian statistics?

Only in an oversimplified manner. You will recall that the classical approach was concerned with calculating the probability (likelihood) of the data given that a preset H_o was true. If this probability was less than α then a rejection of H_o ensued.

In the Bayesian approach the probability of concern is the probability of any H_o being true *given* the data. Notice that the conditional probabilities are reversed.

In symbolic form, the key conditional probabilities are:

Classical: $P(\text{data} \mid H_o \text{ true})$

Bayesian: $P(H_o \text{ true} \mid \text{data})$

The Bayesian argues that if one can calculate the probability of any hypothesis given the data then one can make what is called a decision of *Maximum Rationalization* as to which hypothesis is most likely to be true, namely, the hypothesis with the highest probability *given* the data. This way the Bayesian is not bound to a single H_o as is the Classicist. Bayes' Theorem, which ties together the above probabilities is, in its simplest form,

$$P(H \mid \text{data}) = \frac{P(\text{data} \mid H) P(H)}{P(\text{data})}$$

Note that a part of the Bayes' Theorem is the probability of interest to the Classicist, namely $P(\text{data} \mid H)$. Here $P(H)$ is the probability that H is true.

There are many pros and cons relative to the Classical and Bayesian techniques. Both make good sense if what each does happens to be what you, the researcher, want to do.

How can I decide whether to be a classicist or a Bayesian?

You should not decide since you are, at this stage, in no position intellectually or experience-wise to jump on any bandwagon. To do so now would be acting out of ignorance. Most of the behavioral research published today is classical in nature so you have to know about its philosophy and techniques. As more Bayesian techniques are developed and utilized in behavioral research and other applied areas you will need to become more knowledgeable in that area. You are obligated, as a researcher, to become as informed as possible concerning *any* technique or viewpoint which will aid your understanding and decision making. This is the mark of a good researcher.

SOME STATISTICAL REMINDERS

1. One does not "accept" H_o but rather "fails to reject" H_o. They are not the same.
2. There is no such thing as "almost (nearly, approaching, highly, tends to be, etc.) significant." Each statistical test is significant or not for a particular preset α, β, effect size and sample size.
3. No amount of statistical testing can correct a poorly-designed study.
4. If the assumptions of your statistical test are not met, then it is impossible to tell if the result of your test is valid or interpretable.
5. A nonrandom large sample is no improvement over a nonrandom small sample but a *random* large sample is a vast improvement over a *random* small sample.
6. There are no statistical reasons for not obtaining a sufficiently large random sample; only practical ones (financial, difficulties with subjects, etc.).
7. The calculation of a PPM correlation coefficient does not, "investigate the relationship between variables," but merely gives an indication of the *degree* of linear relationship.

NOTATION

1. μ: Population mean (alternately called arithmetic average, First Moment, Expected Value; sometimes denoted M or E(X)).
2. $\overline{X}$: Sample mean (sometimes denoted $\hat{\mu}, \overline{x}$).
3. σ^2: Population variance (Second Moment about the mean).
4. S^2: Sample variance (sometimes denoted $\hat{\sigma}^2, s^2$).
5. σ: Population standard deviation. It is the positive square root of σ^2.
6. S: Sample standard deviation. It is the positive square root of S^2.
7. ρ: Population Pearson Product-Moment Correlation Coefficient.
8. r: Sample Pearson Product-Moment Correlation Coefficient.
9. r_s: Sample Spearman's Rho Correlation Coefficient.
10. ρ_s: Population Spearman's Rho Coefficient.
11. M_d: Population median (sometimes denoted Mdn).
12. m_d: Sample median.
13. β_0: The population intercept in a simple linear regression equation (sometimes denoted α, A).
14. β_1: The population slope of a simple linear regression equation (sometimes denoted β, B).
15. $\hat{\beta}_0$: Sample estimate of β_0 (sometimes denoted as a, $\hat{\alpha}$).
16. $\hat{\beta}_1$: Sample estimate of β_1 (sometimes denoted b, $\hat{\beta}$).
17. C.I.: Confidence Interval.
18. α: Probability of rejecting H_o when H_o is true.
19. β: Probability of failing to reject H_o when H_o is false.
20. Power: $1 - \beta$ = Probability of rejecting H_o when H_o is false.
21. P: Population proportion possessing a certain characteristic (for example, proportion of males in a population).

22. p: Sample proportion. An estimate of P using a sample. (Sometimes denoted $\hat{P}$.) Also used to denote the probability associated with the data given H_o is true.
23. P(X): Probability of X (sometimes denoted as Pr(X), p(X)). See Appendix C.
24. a > b: a greater than b.
25. a < b: a less than b.
26. a ≤ b: a less than or equal to b (b greater than or equal to a).
27. n: sample size.

Table A
t_α values*

	α Level for one-tailed test					
	.10	.05	.025	.01	.005	.0005
df	α Level for two-tailed test					
	.20	.10	.05	.02	.01	.001
1	3.078	6.314	12.706	31.821	63.657	636.619
2	1.886	2.920	4.303	6.965	9.925	31.598
3	1.638	2.353	3.182	4.541	5.841	12.941
4	1.533	2.132	2.776	3.747	4.604	8.610
5	1.476	2.015	2.571	3.365	4.032	6.859
6	1.440	1.943	2.447	3.143	3.707	5.959
7	1.415	1.895	2.365	2.998	3.499	5.405
8	1.397	1.860	2.306	2.896	3.355	5.041
9	1.383	1.833	2.262	2.821	3.250	4.781
10	1.372	1.812	2.228	2.764	3.169	4.587
11	1.363	1.796	2.201	2.718	3.106	4.437
12	1.356	1.782	2.179	2.681	3.055	4.318
13	1.350	1.771	2.160	2.650	3.012	4.221
14	1.345	1.761	2.145	2.624	2.977	4.140
15	1.341	1.753	2.131	2.602	2.947	4.073
16	1.337	1.746	2.120	2.583	2.921	4.015
17	1.333	1.740	2.110	2.567	2.898	3.965
18	1.330	1.734	2.101	2.552	2.878	3.922
19	1.328	1.729	2.093	2.539	2.861	3.883
20	1.325	1.725	2.086	2.528	2.845	3.850
21	1.323	1.721	2.080	2.518	2.831	3.819
22	1.321	1.717	2.074	2.508	2.819	3.792
23	1.319	1.714	2.069	2.500	2.807	3.767
24	1.318	1.711	2.064	2.492	2.797	3.745
25	1.316	1.708	2.060	2.485	2.787	3.725
26	1.315	1.706	2.056	2.479	2.779	3.707
27	1.314	1.703	2.052	2.473	2.771	3.690
28	1.313	1.701	2.048	2.467	2.763	3.674
29	1.311	1.699	2.045	2.462	2.756	3.659
30	1.310	1.697	2.042	2.457	2.750	3.646
40	1.303	1.684	2.021	2.423	2.704	3.551
60	1.296	1.671	2.000	2.390	2.660	3.460
120	1.289	1.658	1.980	2.358	2.617	3.373
∞	1.282	1.645	1.960	2.326	2.576	3.291

*Adapted from *Biometrika Tables for Statisticians,* Vol. 1, E.S. Pearson. Cambridge University Press, 1966.

Table B*

F_α values ($\alpha = .05$ one-tailed, $\alpha = .10$ two-tailed)

df_2	\multicolumn{9}{c}{df_1}								
	1	2	3	4	5	6	7	8	9
1	161.4	199.5	215.7	224.6	230.2	234.0	236.8	238.9	240.5
2	18.51	19.00	19.16	19.25	19.30	19.33	19.35	19.37	19.38
3	10.13	9.55	9.28	9.12	9.01	8.94	8.89	8.85	8.81
4	7.71	6.94	6.59	6.39	6.26	6.16	6.09	6.04	6.00
5	6.61	5.79	5.41	5.19	5.05	4.95	4.88	4.82	4.77
6	5.99	5.14	4.76	4.53	4.39	4.28	4.21	4.15	4.10
7	5.59	4.74	4.35	4.12	3.97	3.87	3.79	3.73	3.68
8	5.32	4.46	4.07	3.84	3.69	3.58	3.50	3.44	3.39
9	5.12	4.26	3.86	3.63	3.48	3.37	3.29	3.23	3.18
10	4.96	4.10	3.71	3.48	3.33	3.22	3.14	3.07	3.02
11	4.84	3.98	3.59	3.96	3.20	3.09	3.01	2.95	2.90
12	4.75	3.89	3.49	3.26	3.11	3.00	2.91	2.85	2.80
13	4.67	3.81	3.41	3.18	3.03	2.92	2.83	2.77	2.71
14	4.60	3.74	3.34	3.11	2.96	2.85	2.76	2.70	2.65
15	4.54	3.68	3.29	3.06	2.90	2.79	2.71	2.64	2.59
16	4.49	3.63	3.24	3.01	2.85	2.74	2.66	2.59	2.54
17	4.45	3.59	3.20	2.96	2.81	2.70	2.61	2.55	2.49
18	4.41	3.55	3.16	2.93	2.77	2.66	2.58	2.51	2.46
19	4.38	3.52	3.13	2.90	2.74	2.63	2.54	2.48	2.42
20	4.35	3.49	3.10	2.87	2.71	2.60	2.51	2.45	2.39
21	4.32	3.47	3.07	2.84	2.68	2.57	2.49	2.42	2.37
22	4.30	3.44	3.05	2.82	2.66	2.55	2.46	2.40	2.34
23	4.28	3.42	3.03	2.80	2.64	2.53	2.44	2.37	2.32
24	4.26	3.40	3.01	2.78	2.62	2.51	2.42	2.36	2.30
25	4.24	3.39	2.99	2.76	2.60	2.49	2.40	2.34	2.28
26	4.23	3.37	2.98	2.74	2.59	2.47	2.39	2.32	2.27
27	4.21	3.35	2.96	2.73	2.57	2.46	2.37	2.31	2.25
28	4.20	3.34	2.95	2.71	2.56	2.45	2.36	2.29	2.24
29	4.18	3.33	2.93	2.70	2.55	2.43	2.35	2.28	2.22
30	4.17	3.32	2.92	2.69	2.53	2.42	2.33	2.27	2.21
40	4.08	3.23	2.84	2.61	2.45	2.34	2.25	2.18	2.12
60	4.00	3.15	2.76	2.53	2.37	2.25	2.17	2.10	2.04
120	3.92	3.07	2.68	2.45	2.29	2.17	2.09	2.02	1.96
∞	3.84	3.00	2.60	2.37	2.21	2.10	2.01	1.94	1.88

*Reproduced from Table 18 of *Biometrika Tables for Statisticians*, Vol. 1, E.S. Pearson. Cambridge University Press, 1966.

Table B (Continued)

F_α values ($\alpha = .05$ one-tailed, $\alpha = .10$ two-tailed)

df_2	\multicolumn{10}{c}{df_1}
	10	12	15	20	24	30	40	60	120	∞
1	241.9	243.9	245.9	248.0	249.1	250.1	251.1	252.2	253.3	254.3
2	19.40	19.41	19.43	19.45	19.45	19.46	19.47	19.48	19.49	19.50
3	8.79	8.74	8.70	8.66	8.64	8.62	8.59	8.57	8.55	8.53
4	5.96	5.91	5.86	5.80	5.77	5.75	5.72	5.69	5.66	5.63
5	4.74	4.68	4.62	4.56	4.53	4.50	4.46	4.43	4.40	4.36
6	4.06	4.00	4.94	3.87	3.84	3.81	3.77	3.74	3.70	3.67
7	3.64	3.57	3.51	3.44	3.41	3.38	3.34	3.30	3.27	3.23
8	3.35	3.28	3.22	3.15	3.12	3.08	3.04	3.01	2.97	2.93
9	3.14	3.07	3.01	2.94	2.90	2.86	2.83	2.79	2.75	2.71
10	2.98	2.91	2.85	2.77	2.74	2.70	2.66	2.62	2.58	2.54
11	2.85	2.79	2.72	2.65	2.61	2.57	2.53	2.49	2.45	2.40
12	2.75	2.69	2.62	2.54	2.51	2.47	2.43	2.38	2.34	2.30
13	2.67	2.60	2.53	2.46	2.42	2.38	2.34	2.30	2.25	2.21
14	2.60	2.53	2.46	2.39	2.35	2.31	2.27	2.22	2.18	2.13
15	2.54	2.48	2.40	2.33	2.29	2.25	2.20	2.16	2.11	2.07
16	2.49	2.42	2.35	2.28	2.24	2.19	2.15	2.11	2.06	2.01
17	2.45	2.38	2.31	2.23	2.19	2.15	2.10	2.06	2.01	1.96
18	2.41	2.34	2.27	2.19	2.15	2.11	2.06	2.02	1.97	1.92
19	2.38	2.31	2.23	2.16	2.11	2.07	2.03	1.98	1.93	1.88
20	2.35	2.28	2.20	2.12	2.08	2.04	1.99	1.95	1.90	1.84
21	2.32	2.25	2.18	2.10	2.05	2.01	1.96	1.92	1.87	1.81
22	2.30	2.23	2.15	2.07	2.03	1.98	1.94	1.89	1.84	1.78.
23	2.27	2.20	2.13	2.05	2.01	1.96	1.91	1.86	1.81	1.76
24	2.25	2.18	2.11	2.03	1.98	1.94	1.89	1.84	1.79	1.73
25	2.24	2.16	2.09	2.01	1.96	1.92	1.87	18.2	1.77	1.71
26	2.22	2.15	2.07	1.99	1.95	1.90	1.85	1.80	1.75	1.69
27	2.20	2.13	2.06	1.97	1.93	1.88	1.84	1.79	1.73	1.67
28	2.19	2.12	2.04	1.96	1.91	1.87	1.82	1.77	1.71	1.65
29	2.18	2.10	2.03	1.94	1.90	1.85	1.81	1.75	1.70	1.64
30	2.16	2.09	2.01	1.93	1.89	1.84	1.79	1.74	1.68	1.62
40	2.08	2.00	1.92	1.84	1.79	1.74	1.69	1.64	1.58	1.51
60	1.99	1.92	1.84	1.75	1.70	1.65	1.59	1.53	1.47	1.39
120	1.91	1.83	1.75	1.66	1.61	1.55	1.50	1.43	1.35	1.25
∞	1.83	1.75	1.67	1.57	1.52	1.46	1.39	1.32	1.22	1.00

Table B (Continued)

F_α values ($\alpha = .01$ one-tailed, $\alpha = .02$ two-tailed)

					df_1				
df_2	1	2	3	4	5	6	7	8	9
1	4052	4999.5	5403	5625	5764	5859	5928	5981	6022
2	98.50	99.00	99.17	99.25	99.30	99.33	99.36	99.37	99.39
3	34.12	30.82	29.46	28.71	28.24	27.91	27.67	27.49	27.35
4	21.20	18.00	16.69	15.98	15.52	15.21	14.98	14.80	14.66
5	16.26	13.27	12.06	11.39	10.97	10.67	10.46	10.29	10.16
6	13.75	10.92	9.78	9.15	8.75	8.47	8.26	8.10	7.98
7	12.25	9.55	8.45	7.85	7.46	7.19	6.99	6.84	6.72
8	11.26	8.65	7.59	7.01	6.63	6.37	6.18	6.03	5.91
9	10.56	8.02	6.99	6.42	6.06	5.80	5.61	5.47	5.35
10	10.04	7.56	6.55	5.99	5.64	5.39	5.20	5.06	4.94
11	9.65	7.21	6.22	5.67	5.32	5.07	4.89	4.74	4.63
12	9.33	6.93	5.95	5.41	5.06	4.82	4.64	4.50	4.39
13	9.07	6.70	5.74	5.21	4.86	4.62	4.44	4.30	4.19
14	8.86	6.51	5.56	5.04	4.69	4.46	4.28	4.14	4.03
15	8.68	6.36	5.42	4.89	4.56	4.32	4.14	4.00	3.89
16	8.53	6.23	5.29	4.77	4.44	4.20	4.03	3.89	3.78
17	8.40	6.11	5.18	4.67	4.34	4.10	3.93	3.79	3.68
18	8.29	6.01	5.09	4.58	4.25	4.01	3.84	3.71	3.60
19	8.18	5.93	5.01	4.50	4.17	3.94	3.77	3.63	3.52
20	8.10	5.85	4.94	4.43	4.10	3.87	3.70	3.56	3.46
21	8.02	5.78	4.87	4.37	4.04	3.81	3.64	3.51	3.40
22	7.95	5.72	4.82	4.31	3.99	3.76	3.59	3.45	3.35
23	7.88	5.66	4.76	4.26	3.94	3.71	3.54	3.41	3.30
24	7.82	5.61	4.72	4.22	3.90	3.67	3.50	3.36	3.26
25	7.77	5.57	4.68	4.18	3.85	3.63	3.46	3.32	3.22
26	7.72	5.53	4.64	4.14	3.82	3.59	3.42	3.29	3.18
27	7.68	5.49	4.60	4.11	3.78	3.56	3.39	3.26	3.15
28	7.64	5.45	4.57	4.07	3.75	3.53	3.36	3.23	3.12
29	7.60	5.42	4.54	4.04	3.73	3.50	3.33	3.20	3.09
30	7.56	5.39	4.51	4.02	3.70	3.47	3.30	3.17	3.07
40	7.31	5.18	4.31	3.83	3.51	3.29	3.12	2.99	2.89
60	7.08	4.98	4.13	3.65	3.34	3.12	2.95	2.82	2.72
120	6.85	4.79	3.95	3.48	3.17	2.96	2.79	2.66	2.56
∞	6.63	4.61	3.78	3.32	3.02	2.80	2.64	2.51	2.41

Table B (Continued)

F_α values ($\alpha = .01$ one-tailed, $\alpha = .02$ two-tailed)

					df_1					
df_2	10	12	15	20	24	30	40	60	120	∞
1	6056	6106	6157	6209	6235	6261	6287	6313	6339	6366
2	99.40	99.42	99.43	99.45	99.46	99.47	99.47	99.48	99.49	99.50
3	27.23	27.05	26.87	26.69	26.60	26.50	26.41	26.32	26.22	26.13
4	14.55	14.37	14.20	14.02	13.93	13.84	13.75	13.65	13.56	13.46
5	10.05	9.89	9.72	9.55	9.47	9.38	9.29	9.20	9.11	9.02
6	7.87	7.72	7.56	7.40	7.31	7.23	7.14	7.06	6.97	6.88
7	6.62	6.47	6.31	6.16	6.07	5.99	5.91	5.82	5.74	5.65
8	5.81	5.67	5.52	5.36	5.28	5.20	5.12	5.03	4.95	4.86
9	5.26	5.11	4.96	4.81	4.73	4.65	4.57	4.48	4.40	4.31
10	4.85	4.71	4.56	4.41	4.33	4.25	4.17	4.08	4.00	3.91
11	4.54	4.40	4.25	4.10	4.02	3.94	3.86	3.78	3.69	3.60
12	4.30	4.16	4.01	3.86	3.78	3.70	3.62	3.54	3.45	3.36
13	4.10	3.96	3.82	3.66	3.59	3.51	3.43	3.34	3.25	3.17
14	3.94	3.80	3.66	3.51	3.43	3.35	3.27	3.18	3.09	3.00
15	3.80	3.67	3.52	3.37	3.29	3.21	3.13	3.05	2.96	2.87
16	3.69	3.55	3.41	3.26	3.18	3.10	3.02	2.93	2.84	2.75
17	3.59	3.46	3.31	3.16	3.08	3.00	2.92	2.83	2.75	2.65
18	3.51	3.37	3.23	3.08	3.00	2.92	2.84	2.75	2.66	2.57
19	3.43	3.30	3.15	3.00	2.92	2.84	2.76	2.67	2.58	2.49
20	3.37	3.23	3.09	2.94	2.86	2.78	2.69	2.61	2.52	2.42
21	3.31	3.17	3.03	2.88	2.80	2.72	2.64	2.55	2.46	2.36
22	3.26	3.12	2.98	2.83	2.75	2.67	2.58	2.50	2.40	2.31
23	3.21	3.07	2.93	2.78	2.70	2.62	2.54	2.45	2.35	2.26
24	3.17	3.03	2.89	2.74	2.66	2.58	2.49	2.40	2.31	2.21
25	3.13	2.99	2.85	2.70	2.62	2.54	2.45	2.36	2.27	2.17
26	3.09	2.96	2.81	2.66	2.58	2.50	2.42	2.33	2.23	2.13
27	3.06	2.93	2.78	2.63	2.55	2.47	2.38	2.29	2.20	2.10
28	3.03	2.90	2.75	2.60	2.52	2.44	2.35	2.26	2.17	2.06
29	3.00	2.87	2.73	2.57	2.49	2.41	2.33	2.23	2.14	2.03
30	2.98	2.84	2.70	2.55	2.47	2.39	2.30	2.21	2.11	2.01
40	2.80	2.66	2.52	2.37	2.29	2.20	2.11	2.02	1.92	1.80
60	2.63	2.50	2.35	2.20	2.12	2.03	1.94	1.84	1.73	1.60
120	2.47	2.34	2.19	2.03	1.95	1.86	1.76	1.66	1.53	1.38
∞	2.32	2.18	2.04	1.88	1.79	1.70	1.59	1.47	1.32	1.00

Table C*

X_α^2 Values (two-tailed)

df	$a = .995$	$a = .99$	$a = .975$	$a = .95$	$a = .05$	$a = .025$	$a = .01$	$a = .005$	df
1	.0000393	.000157	.000982	.00393	3.841	5.024	6.635	7.879	1
2	.0100	.0201	.0506	.103	5.991	7.378	9.210	10.597	2
3	.0717	.115	.216	.352	7.815	9.348	11.345	12.838	3
4	.207	.297	.484	.711	9.488	11.143	13.277	14.860	4
5	.412	.554	.831	1.145	11.070	12.832	15.086	16.750	5
6	.676	.872	1.237	1.635	12.592	14.449	16.812	18.548	6
7	.989	1.239	1.690	2.167	14.067	16.013	18.475	20.278	7
8	1.344	1.646	2.180	2.733	15.507	17.535	20.090	21.955	8
9	1.735	2.088	2.700	3.325	16.919	19.023	21.666	23.589	9
10	2.156	2.558	3.247	3.940	18.307	20.483	23.209	25.188	10
11	2.603	3.053	3.816	4.575	19.675	21.920	24.725	26.757	11
12	3.074	3.571	4.404	5.226	21.026	23.337	26.217	28.300	12
13	3.565	4.107	5.009	5.892	22.362	24.736	27.688	29.819	13
14	4.075	4.660	5.629	6.571	23.685	26.119	29.141	31.319	14
15	4.601	5.229	6.262	7.261	24.996	27.488	30.578	32.801	15
16	5.142	5.812	6.908	7.962	26.296	28.845	32.000	34.267	16
17	5.697	6.408	7.564	8.672	27.587	30.191	33.409	35.718	17
18	6.265	7.015	8.231	9.390	28.869	31.526	34.805	37.156	18
19	6.844	7.633	8.907	10.117	30.144	32.852	36.191	38.582	19
20	7.434	8.260	9.591	10.851	31.410	34.170	37.566	39.997	20
21	8.034	8.897	10.283	11.591	32.671	35.479	38.932	41.401	21
22	8.643	9.542	10.982	12.338	33.924	36.781	40.289	42.796	22
23	9.260	10.196	11.689	13.091	35.172	38.076	41.638	44.181	23
24	9.886	10.856	12.401	13.848	36.415	39.364	42.980	45.558	24
25	10.520	11.524	13.120	14.611	37.652	40.646	44.314	46.928	25
26	11.160	12.198	13.844	15.379	38.885	41.923	45.642	48.290	26
27	11.808	12.879	14.573	16.151	40.113	43.194	46.963	49.645	27
28	12.461	13.565	15.308	16.928	41.337	44.461	48.278	50.993	28
29	13.121	14.256	16.047	17.708	42.557	45.722	49.588	52.336	29
30	13.787	14.953	16.791	18.493	43.773	46.979	50.892	53.672	30

*Adapted from *Biometrika Tables for Statisticians,* Vol. 1, E.S. Pearson. Cambridge University Press, 1966.

Table D

Normal Curve Areas:

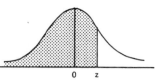

z	.00	.01	.02	.03	.04	.05	.06	.07	.08	.09
.0	.5000	.5040	.5080	.5120	.5160	.5199	.5239	.5279	.5319	.5359
.1	.5398	.5438	.5478	.5517	.5557	.5596	.5636	.5675	.5714	.5753
.2	.5793	.5832	.5871	.5910	.5948	.5987	.6026	.6064	.6103	.6141
.3	.6179	.6217	.6255	.6293	.6331	.6368	.6406	.6443	.6480	.6517
.4	.6554	.6591	.6628	.6664	.6700	.6736	.6772	.6808	.6844	.6879
.5	.6915	.6950	.6985	.7019	.7054	.7088	.7123	.7157	.7190	.7224
.6	.7257	.7291	.7324	.7357	.7389	.7422	.7454	.7486	.7517	.7549
.7	.7580	.7611	.7642	.7673	.7704	.7734	.7764	.7794	.7823	.7852
.8	.7881	.7910	.7939	.7967	.7995	.8023	.8051	.8078	.8106	.8133
.9	.8159	.8186	.8212	.8238	.8264	.8289	.8315	.8340	.8365	.8389
1.0	.8413	.8438	.8461	.8485	.8508	.8531	.8554	.8577	.8599	.8621
1.1	.8643	.8665	.8686	.8708	.8729	.8749	.8770	.8790	.8810	.8830
1.2	.8849	.8869	.8888	.8907	.8925	.8944	.8962	.8980	.8997	.9015
1.3	.9032	.9049	.9066	.9082	.9099	.9115	.9131	.9147	.9162	.9177
1.4	.9192	.9207	.9222	.9236	.9251	.9265	.9279	.9292	.9306	.9319
1.5	.9332	.9345	.9357	.9370	.9382	.9394	.9406	.9418	.9429	.9441
1.6	.9452	.9463	.9474	.9484	.9495	.9505	.9515	.9525	.9535	.9545
1.7	.9554	.9564	.9573	.9582	.9591	.9599	.9608	.9616	.9625	.9633
1.8	.9641	.9649	.9656	.9664	.9671	.9678	.9686	.9693	.9699	.9706
1.9	.9713	.9719	.9726	.9732	.9738	.9744	.9750	.9756	.9761	.9767
2.0	.9772	.9778	.9783	.9788	.9793	.9798	.9803	.9808	.9812	.9817
2.1	.9821	.9826	.9830	.9834	.9838	.9842	.9846	.9850	.9854	.9857
2.2	.9861	.9864	.9868	.9871	.9875	.9878	.9881	.9884	.9887	.9890
2.3	.9893	.9896	.9898	.9901	.9904	.9906	.9909	.9911	.9913	.9916
2.4	.9918	.9920	.9922	.9925	.9927	.9929	.9931	.9932	.9934	.9936
2.5	.9938	.9940	.9941	.9943	.9945	.9946	.9948	.9949	.9951	.9952
2.6	.9953	.9955	.9956	.9957	.9959	.9960	.9961	.9962	.9963	.9964
2.7	.9965	.9966	.9967	.9968	.9969	.9970	.9971	.9972	.9973	.9974
2.8	.9974	.9975	.9976	.9977	.9977	.9978	.9979	.9979	.9980	.9981
2.9	.9981	.9982	.9982	.9983	.9984	.9984	.9985	.9985	.9986	.9986
3.0	.9987	.9987	.9987	.9988	.9988	.9989	.9989	.9989	.9990	.9990
3.1	.9990	.9991	.9991	.9991	.9992	.9992	.9992	.9992	.9993	.9993
3.2	.9993	.9993	.9994	.9994	.9994	.9994	.9994	.9995	.9995	.9995
3.3	.9995	.9995	.9995	.9996	.9996	.9996	.9996	.9996	.9996	.9997
3.4	.9997	.9997	.9997	.9997	.9997	.9997	.9997	.9997	.9997	.9998

*Adapted from Table 1 of *Biometrika Tables for Statisticians*, Vol. 1, E.S. Pearson. Cambridge University Press, 1966.

Appendix A

BASIC MATHEMATICS SKILLS REFRESHER

The purpose of this appendix is to refresh the student in basic mathematics skills, the mastery of which is crucial to obtaining any reasonable level of proficiency in statistical methodology. The reason these skills are crucial is that, almost daily, these skills are brought to bear on methodological problems in statistics. The student's traumas of a mathematical nature will be minimized with mastery of these skills. This frees the student's mind to be concerned with more important aspects of statistical methodology, namely, the "why" and "so what" of the inferences and descriptions being made. The necessary objectives for mastery of basic mathematics skills will be presented below with rules, example problems, exercises and answers where appropriate. The student should study the rules and examples, work the exercises, and make up several problems similar to the examples given. A review test, with answers, will be given at the end of the objectives.

Objective 1: Given an expression involving several arithmetic operations (with or without grouping signs), the student will be able to carry out the indicated operations in the correct order and arrive at a correct answer.

Rule I: In multiplying or dividing, if the numbers have the same sign then the answer is positive (+) in sign. If the signs are different the answer is negative (−) in sign.

Examples:

1. $10 \div 5 = 2$
2. $-10 \div -5 = 2$
3. $-10 \div 5 = -2$
4. $10 \div -5 = -2$
5. $(10)(-5) = -50$
6. $(-10)(-5) = 50$

Exercises:

1. $-7(5) =$ _____

2. $(-7)(5) =$ _____

3. $9 \div -3 =$ _____

Answers:

1. -35
2. -35
3. -3

Rule II: In adding two signed numbers, if the signs are the same then the answer is the sum of the two numbers with the common sign attached. If the signs are different then the answer is the difference between the numbers with the sign of the larger number attached.

Examples:

1. $7 + 3 = 10$
2. $7 + (-3) = 7 - 3 = 4$
3. $-7 - 3 = -10$
4. $-7 + 3 = -4$

Exercises:

1. $-8 + (-5) =$ _____

2. $-8 + 5 =$ _____

3. $7 + (-5) =$ _____

Answers:

1. -13
2. -3
3. 2

Rule III: In subtracting any number from another, change the sign of the number being subtracted and add the two numbers.

Examples:

1. Subtracting 3 from 7 is 7 + (−3) = 4
2. Subtracting 7 from 3 is 3 + (−7) = −4
3. Subtracting −3 from 7 is 7 + (+3) = 10

Note: −(−3) is the same as saying "subtract minus 3." So subtracting −3 from 7 can be written 7 − (−3) = 7 + 3 = 10

Exercises:

1. −6 subtracted from −2 =
2. 3 subtracted from −2 =

Answers:

1. 4
2. −5

Rule IV:
a. *If the multiplication, division, addition, and subtraction operations are mixed in an expression, then multiplication and division are conducted first followed by addition and subtraction. If parentheses are present, then those operations inside the parentheses are conducted first.*
b. $a + b = b + a$
c. $(a + b) + c = a + (b + c)$
d. $a(b + c) = ab + ac$

Examples:

1. 3 − 6 · 2 = 3 − 12 = −9
2. 7 + 6 ÷ 3 = 7 + 2 = 9
3. 4 + 7 − 3 = 8
4. (3 − 6)2 = −3(2) = −6
5. (9 − 3) ÷ 2 = 6 ÷ 2 = 3
6. 2(3 + 7) = 2 · 3 + 2 · 7 = 6 + 14
7. (6 + 3) + 9 = 9 + 9 = 18
8. 6 + (3 + 9) = 6 + 12 = 18

Exercises:

1. (8 + 6 − 2) ÷ 2 =
2. 8 + 6 − 2 ÷ 2 =
3. (7 + 5)3 =
4. 7 + 5 · 3 =

Answers:

1. 6
2. 13
3. 36
4. 22

Objective 2: The student will be able to correctly carry out arithmetic operations with signed expressions containing fractions, decimals and exponents.

Rule I: To multiply two fractions, multiply the numerators together and the denominators together.

Example:

1. $(1/2)(3/4) = \dfrac{(1)(3)}{(2)(4)} = 3/8$
2. $(-7/8)(2/3) = -14/24 = 14/-24 = -7/12$
3. $(-7/8)(-2/3) = 14/24 = 7/12$

Note: A negative in front of a fraction means either the denominator *or* the numerator is negative but *not* both. (If both were negative the fraction would be positive).

Exercises:

1. $(-3/4)(2/3) =$ _____

2. $-(5/2)(3/5) =$ _____

Answers:

1. $-6/12 = -1/2$
2. $-15/10 = -3/2$

Rule II: To divide by a fraction, invert the fraction by which you are dividing [divisor] then multiply.

Examples:

1. $1/2 \div 2/3 = (1/2)(3/2) = 3/4$
2. $(1/2 \div (-7/8) = (1/2)(-8/7) = -8/14 = -4/7$
3. $-1/2/-7/8 = (-1/2)(-8/7) = 8/14 = 4/7$

Exercises:

1. $7/5 \div 2/3 =$ _____

2. −3/8 ÷ 7/9 = _____

3. (4/5)/(3/5) = _____

Answers:

1. 21/10
2. −27/56
3. 4/3

Rule III: To add or subtract fractions, find a common denominator and then add or subtract the numerators and retain the common denominator.

Examples:

1. 1/2 + 2/3 = (1/2)(3/3) + (2/3)(2/2) = 3/6 + 4/6 = 7/6
2. 1/2 − 2/3 = 3/6 − 4/6 = −1/6

Exercises:

1. 3/4 − 2/3 = _____

2. −5/2 − 7/6 = _____

Answers:

1. 1/12
2. −22/6

Rule IV: In multiplying decimal numbers, multiply the numbers as if there were no decimals then add the number of total decimal places used and count this number of places from the right to find the decimal place for the answer.

Examples:

1. (.3)(.4) = .12
2. (.03)(−.4) = −.012
3. (2.5)(.002) = .0050

Exercises:

1. (.2)(.03) = _____

2. (−.04)(2.3) = _____

3. (−.002)(−3.2) = _____

Answers:

1. .006
2. −.092
3. .0064

Rule V: To divide a decimal by a decimal number, move the decimal point in the numerator and the denominator the same number of places so that the denominator and numerator are whole numbers, then divide out or reduce to a simpler fraction.

Examples:

1. .3/.08 = 30./8. = 15/4 = 3.75
2. .025/.50 = 25/500 = 1/20 = .05
3. −2.6/.05 = −260/5 = −52

Exercises:

1. .72/3.2 = _____

2. −2.8/−.0014 = _____

3. .003/.15 = _____

Answers:

1. 9/40 = .225
2. 2000
3. 1/50 = .02

Rule VI:
a. In multiplication of expressions involving exponents with the same base, the base is retained and the exponents are added. In division, the base (being the same) is retained and the exponent of the divisor (denominator) is subtracted from the exponent of the dividend (numerator).
b. If an expression involving exponents is raised to a power then the power is multiplied by the exponents and the base retained.

Examples:

1. $(x^3)(x^5) = x^8$
2. $2x^5/x^3 = 2x^2$
3. $x^7 \div x^6 = x$

4. $(3x^2)(y^3) = 3x^2y^3$
5. $(6x^3)^2 = 6^2x^6 = 36x^6$
6. $(-2y^2)^2 = (-2)^2y^4 = 4y^4$

Exercises:

1. $(3x^5)(x^3) = $ _____

2. $-x^3/x^2 = $ _____

3. $6x^5 \div 3x^2 = $ _____

4. $(5x^3)^2 = $ _____

Answers:

1. $3x^8$
2. $-x$
3. $2x^3$
4. $25x^6$

Rule VII: In adding and subtracting expressions involving exponents, add and subtract only those expressions which have identical bases and exponents.

Examples:

1. $3x^2 + 7x^2 = 10x^2$
2. $3y^2 + 7x^2 = 3y^2 + 7x^2$
3. $-3y^5 - 2y^5 = -5y^5$

Exercise:

1. $2x^3 + 7x^2 = $ _____

2. $3x^5 + 7x^5 = $ _____

3. $-5x^2 + 6x^2 = $ _____

Answers:

1. $2x^3 + 7x^2$
2. $10x^5$
3. x^2

Objective 3: Given an expression involving square roots, the student will be able to perform the necessary operations and affix the proper decimal point, if necessary.

Rule I:
a. The square root of a fraction is the fraction of square roots.
b. The square root of a product is the product of the square roots.

Examples:

1. $\sqrt{5/3} = \sqrt{5}/\sqrt{3}$
2. $\sqrt{(7)(6)} = \sqrt{7} \cdot \sqrt{6} = \sqrt{42}$
3. $\sqrt{a} \cdot \sqrt{b} = \sqrt{ab}$

Exercises:

1. $\sqrt{16/25} = $ _____

2. $2\sqrt{3}\sqrt{6} = $ _____

Answers:

1. 4/5
2. $2\sqrt{18} = 2\sqrt{(9)(2)} = 2\sqrt{9}\sqrt{2} = (2)(3)\sqrt{2} = 6\sqrt{2}$

Rule II: If a square root cannot be looked up directly in the table, then multiply and divide the number under the radical by a power of 100 until the number can be looked up or approximated from a table.

Examples:

1. $\sqrt{.25} = \sqrt{(.25)(100)/100} = \sqrt{25}/\sqrt{100} = 5/10 = 1/2 = .5$
2. $\sqrt{.0144} = \sqrt{(.0144)(10,000/10,000)} = \sqrt{144/10,000} = 12/100 = .12$
3. $\sqrt{12100} = \sqrt{(121)(100)} = \sqrt{121}\sqrt{100} = (11)(10) = 110$

Exercises:

1. $\sqrt{6.25} = $ _____

2. $\sqrt{.0196} = $ _____

3. $\sqrt{49000}$ = _____

　　　　　　　　　　　　　　　Hint: You'll need a table here

Answers:
1. 2.5
2. .14
3. 221.4

Objective 4: Given an expression involving variables and given values for these variables, the student will be able to correctly calculate the value of the expression.

Examples:

1. If $x = 2$ and $y = 3$ then
   $3x^2 - 6y = 3(2^2) - 6(3) = 3(4) - 18 = 12 - 18 = -6$
2. If $x = -1$ and $z = 2$ then
   $3x^3/6z^2 = 3(-1^3)/6(2^2) = 3(-1)/6(4) = -3/24 = -1/8$
3. If $x = 1$ and $y = -2$ then
   $2x^3y^2 + 3y = 2(1^3)(-2^2) + 3(-2) = (2)(4) - 6 = 2$

Exercises:

If $x = 2$ and $y = -3$ then

1. $3x^2y + 2x^3 =$ _____

2. $-7xy^3 \div 3x^2 =$ _____

Answers:
1. -20
2. $63/2$

Objective 5: The student will be able to solve equations with one unknown.

Rule I: Both sides of an equation can have the same number added to or subtracted from them without affecting the solution of the equation. Likewise both sides may be multiplied or divided by the same non-zero number without affecting the equation's solution. (Note: The object in solving equations is to isolate the unknown on one side and everything else on the other side of the equation.)

Examples:

1. Solve for x when
   $3x - 6 = 2$.
   $3x - 6 + 6 = 2 + 6$ (by adding 6 to both sides)
   $3x + 0 = 8$
   $3x/3 = 8/3$ (by dividing both sides by 3)
   $x = 8/3$

2. Solve for x when
   $7x - 4 = 2x + 10$.
   $7x - 4 - 2x = 2x + 10 - 2x$ (by subtracting 2x from both sides)
   $5x - 4 = 10$
   $5x - 4 + 4 = 10 + 4$ (by adding 4 to both sides)
   $\dfrac{5x}{5} = \dfrac{14}{5}$ (by dividing both sides by 5)
   $x = 14/5$

Exercises:

Solve for y if

1. $-7y + 2 = 6y - 3$ _____

2. $6y = 18 + 9y$ _____

Answers:

1. $y = 5/13$
2. $y = -6$

Rule II: If the equation involves a radical, then isolate the radical, square both sides of the equation and solve.

Examples:

1. Solve for x:
   $\sqrt{x} + 6 = 9$.
   $\sqrt{x} + 6 - 6 = 9 - 6$ (Subtracting 6 from both sides)
   $\sqrt{x} = 3$
   $(\sqrt{x})^2 = (3)^2$ (by squaring both sides)
   $x = 9$

2. Solve for y:

$\sqrt{3y - 2} = 7.$
$(\sqrt{3y - 2})^2 = (7)^2$ (by squaring both sides)
$3y - 2 = 49$
$3y - 2 + 2 = 49 + 2$ (by adding 2 to both sides)
$3y = 51$
$3y/3 = 51/3$ (by dividing both sides by 3)
$y = 17$

Exercises:

1. Solve for x:

   $\sqrt{2x} - 3 = 4$ _____

2. Solve for y:

   $-3 = -6 + \sqrt{y - 3}$ _____

Answers:

1. 49/2
2. 12

Objective 6: The student will be able to correctly interpret and calculate with expressions involving summation notation.

The Greek symbol "Σ" is used as a notational device to indicate the formation of a sum. For example, if you had three test scores 2, 3 and 5 and wished to add them then you could label them as

$x_1 = 2$
$x_2 = 3$
$x_3 = 5$

and use the notation

$$\sum_{i=1}^{3} x_i$$

to describe this sum. The notation simply means to "add the three scores together." Thus

$$\sum_{i=1}^{3} x_i = x_1 + x_2 + x_3 = 2 + 3 + 5 = 10.$$

With a great many scores this descriptive notational shorthand is a time and space saver.

Likewise the expression

$$\sum_{i=1}^{3} x^2_i$$ means "square each score and then add," i.e.,

$$\sum_{i=1}^{3} x_i^2 = x_1^2 + x_2^2 + x_3^2 = 2^2 + 3^2 + 5^2 = 38.$$

Note that the "i" subscript is an identification or labeling of the numbers (or scores) to be added. The "i = 1" below the symbol and the "3" above the symbol are there to tell you how many scores are added as well as what is to be added. For example,

$$\sum_{i=1}^{3} ix_i = (1)x_1 + (2)x_2 + (3)x_3 = (1)(2) + (2)(3) + (3)(5) = 23,$$

since the "i" appears as both a subscript *and* as a multiplier. The value for "i" can begin and end at any number depending on the problem at hand. For instance, using the scores $x_1 = 2$, $x_2 = 3$, $x_3 = 5$ and wishing to add the last two scores only for some reason, the notation

$$\sum_{i=2}^{3} x_i$$

would describe what you wish to do since

$$\sum_{i=2}^{3} x_i = x_2 + x_3 = 3 + 5 = 8$$

Rule I: If a constant is a common factor inside a summation symbol it may be factored outside the symbol.

Examples:

1. $\sum_{i=1}^{3} 5x_i = 5\sum_{i=1}^{3} x_i = 5(x_1 + x_2 + x_3)$

2. $\sum_{i=1}^{n} 10y_i^2 = 10 \sum_{i=1}^{n} y_i^2 = 10(y_1^2 + y_2^2 + \ldots + y_n^2)$

3. $\sum_{i=1}^{6} ax_i y_i = a \sum_{i=1}^{6} x_i y_i = a(x_1 y_1 + x_2 y_2 + x_3 y_3 + x_4 y_4 + x_5 y_5 + x_6 y_6)$

Exercises:

1. $\sum_{i=1}^{2} 3 \cdot y_i^2 = $ _____

2. $\sum_{i=1}^{n} (a + b)x_i = $ _____

Answers:

1. $3\sum_{i=1}^{2} y_i^2 = 3(y_1^2 + y_2^2)$

2. $(a + b) \sum_{i=1}^{n} x_i = (a + b)(x_1 + x_2 + \ldots + x_n)$

Rule II: The sum of a sum of expressions is the sum of the separate expressions.

Examples:

1. $\sum_{i=1}^{3} (x_i + y_i) = \sum_{i=1}^{3} x_i + \sum_{i=1}^{3} y_i$

2. $\sum_{i=1}^{n} (3x_i + 2y_i - 7x_i y_i) =$

   $3\sum_{i=1}^{n} x_i + 2\sum_{i=1}^{n} y_i - 7\sum_{i=1}^{n} x_i y_i$

3. $\sum_{i=1}^{5} x_i + 2 = x_1 + x_2 + x_3 + x_4 + x_5 + 2$

4. $\sum_{i=1}^{5} (x_i + 2) = \sum_{i=1}^{5} x_i + \sum_{i=1}^{5} 2 = x_1 + x_2 + x_3 + x_4 + x_5 + 10$

5. $\sum_{i=1}^{3} (x_i - 3)^2 = (x_1 - 3)^2 + (x_2 - 3)^2 + (x_3 - 3)^2$

Exercises:

1. $\sum_{j=1}^{n} (2x_j - 3y_j) = $ _____

2. $\sum_{i=1}^{3} (7x_i^2 + 2x_i) = $ _____

3. $\sum_{i=1}^{3} (x_i - 2) = $ _____

4. $\sum_{i=1}^{3} x_i - 2 = $ _____

5. $\sum_{i=1}^{2} (x_i + 6)^2 = $ _____

Hint: You may add each term then square or square first then add.

Answers:

1. $2\sum_{j=1}^{n} x_j - 3\sum_{j=1}^{n} y_j$

2. $7\sum_{i=1}^{3} x_i^2 + 2\sum_{i=1}^{3} x_i$

3. $x_1 + x_2 + x_3 - 6$

4. $x_1 + x_2 + x_3 - 2$

5. $(x_1 + 6)^2 + (x_2 + 6)^2$ or $\sum_{i=1}^{2} x_i^2 + 12 \sum_{i=1}^{2} x_i + 72$

(If you square first then add)

Rule III: If a constant is the only expression inside a summation symbol then the sum is the product of the expression and the range of the sum.

Examples:

1. $\sum_{i=1}^{3} 4 = 4 + 4 + 4 = (3)(4) = 12$

2. $\sum_{i=1}^{n} (a + b) = n(a + b)$

3. $\sum_{j=1}^{5} C = 5 \cdot C$

4. $\sum_{i=2}^{7} 3 = 6 \cdot 3 = 18$ (Note: Only 6 threes were added)

Exercises:

1. $\sum_{i=1}^{5} 3 =$ _____

2. $\sum_{i=3}^{5} C =$ _____

Answers:

1. 15
2. 3C

BASIC MATH SKILLS
Review Test

Perform the operations indicated

1. $2 + 6 \div 2 =$ _____

2. $[2(3 + 3)^2] + 4 - 3 =$ _____

3. $3[(100 - 40) \div 20] + 6 =$ _____

4. $1/4 \cdot 6/7 =$ _____

5. $3/4 \div 3/2 =$ _____

6. $2/3 + 1/3 =$ _____

7. $1/5 - 3/4 =$ _____

8. $3/4 + 2/3 =$ _____

9. $4/9 - 5/6 =$ _____

10. $24/5 \div 6 =$ _____

11. $\dfrac{3/5}{7} =$ _____

12. $(.3)(.5) =$ _____

13. $(.06)(.423) =$ _____

14. $.24/.006 =$ _____

15. $\sqrt{.49} =$ _____

16. $9 - 13 =$ _____

17. $(-2)(-6) =$ _____

18. $(-5)(2) =$ _____

19. +18/−9 = _____

20. −35/−7 = _____

21. 3(x + 2) = _____

22. $\sqrt{49x^4y^2}$ = _____

23. $(x^2)(x)$ = _____

24. $(3x^3)^2$ = _____

25. $\sqrt{1.96}$ = _____

Express the following as a single radical or a whole number.

26. $\sqrt{3}\sqrt{4}$ = _____

27. $\sqrt{27}/\sqrt{9}$ = _____

28. $4\sqrt{2}$ = _____

29. $\sqrt{16} + \sqrt{9}$ = _____

30. $\sqrt{2}\sqrt{2}$ = _____

31. If y = 3x + 2 and x = 3 then y = _____

32. If y = 2, x = 3 then $2x^2y^3$ = _____

Solve the following for x

33. x + 3 = 15   x = _____

34. 5x − 3 = 17   x = _____

35. 5x − 12 = 6x + 4   x = _____

36. $\sqrt{x − 3}$ = 6   x = _____

Evaluate the following when $x_1 = 3, x_2 = 4, x_3 = 2$

37. $\sum_{i=1}^{3} x_i = $ _____

38. $\sum_{i=2}^{3} 3x_i = $ _____

39. $\sum_{i=1}^{3} 5 = $ _____

40. $\sum_{i=1}^{3} x_i^2 = $ _____

41. $[\sum_{i=1}^{3} x_i]^2 = $ _____

42. $\sum_{i=1}^{3} x_i - 1 = $ _____

43. $\sum_{i=1}^{3} (x_i - 1) = $ _____

44. $\sum_{i=1}^{3} (x_i - 1)^2 = $ _____

BASIC MATH SKILLS
Review Test Answers

1. 5
2. 73
3. 15
4. 3/14
5. 1/2
6. 1
7. −11/20
8. 17/12
9. −7/18
10. 4/5
11. 3/35
12. .15
13. .02538
14. 40
15. .7
16. −4
17. 12
18. −10
19. −2
20. 5
21. $3x + 6$
22. $7x^2y$
23. x^3
24. $9x^6$
25. 1.4
26. $\sqrt{12}$
27. $\sqrt{3}$
28. $\sqrt{32}$
29. 7 or $\sqrt{49}$
30. 2 or $\sqrt{4}$
31. 11
32. 144
33. 12
34. 4
35. −16
36. 39
37. 9
38. 18
39. 15
40. 29
41. 81
42. 8
43. 6
44. 14

Appendix B

FINDING THE SQUARE ROOT

The expression "$\sqrt{x}$" simply means, "find a number (denoted $\sqrt{x}$) such that the number squared (multiplied by itself) is x." For example, $\sqrt{4}$ is a number such that when it is squared the answer will be 4. Now you know that 2 will work since $2^2 = 2 \cdot 2 = 4$. Obviously, $\sqrt{4} \sqrt{4} = 4$ since $\sqrt{4} = 2$. So any "square root," e.g. $\sqrt{x}$, squared is equal to the expression under the symbol "$\sqrt{}$." (The symbol is called a "radical"—God knows why.)

Numbers like 4, 16, 25, for example, have easy square roots since 4, 16, 25, are squares (perfect squares) of well-known whole numbers, namely 2, 4, and 5. However, there will be times when you will need to find the square root of a number which is not as obvious or as well-known.

Let us assume that you have a table of squares and square roots for all the whole numbers from 1 to 1000. Obviously, if the number for which you wish to find the square root is a whole number between 1 and 1000, then the simplest thing to do is look it up. If, however, the number is not a whole number between 1 and 1000, you can resort to one of the following two techniques.

A. Convert the number to a well-known perfect square, or a number in your table, by factoring or multiplying and dividing the number by a power of 100 under the radical until the number remaining is in a usable form (perfect square or between 1 and 1000).

Examples:

1. $\sqrt{1.44} = \sqrt{\frac{(1.44)(100)}{100}} = \sqrt{\frac{144}{100}} = \frac{\sqrt{144}}{\sqrt{100}} = \frac{12}{10} = 1.2$

2. $\sqrt{.0625} = \sqrt{\frac{(.0625)(10,000)}{10,000}} = \sqrt{\frac{625}{10,000}} = \frac{\sqrt{625}}{\sqrt{10,000}} = \frac{25}{100} = .25$

3. $\sqrt{22500} = \sqrt{(225)(100)} = \sqrt{225} \sqrt{100} = (15)(10) = 150$

Exercises:

1. $\sqrt{1.96}$
2. $\sqrt{8100}$
3. $\sqrt{.0121}$ =

Answers:

1. 1.4
2. 90
3. .11

B. If the number cannot be converted to a whole number between 1 and 1000 by technique (A), then approximations must be made. This technique involves simply making repeated intelligent guesses. The steps are as follows:
   1. Convert (using technique A) the number to a nonwhole number between 1 and 1000.
   2. Find the numbers above and below it in the table.
   3. Look up their square roots.
   4. Guess where your square root should be.
   5. Check your guess by squaring it.
   6. If it's close enough for your purposes then keep this first guess.
   7. If further accuracy is needed, let your second guess be halfway between your first guess and
      a. The square root table value larger than your number if your guess was too small or
      b. The square root table value smaller than your number if your guess was too large.

(Hint: If your first guess is not close enough, a slight "common sense" adjustment in the proper direction will often be quite accurate.)

Examples:

1. Find $\sqrt{12.3}$

Now, you know that 12.3 is between 12 and 13 and is closer to 12 than 13. In fact, 12.3 is .3 of the distance between 12 and 13. So a good, common sense guess at the answer would be .3 of the distance between $\sqrt{12}$ and $\sqrt{13}$ both of which are in your table. Consulting the table we find that

$\sqrt{12}$ = 3.4641
$\sqrt{13}$ = 3.6056

This gives a difference of .1415. Now (.3)(.1415) = .04245. Adding this value to the smaller square root gives

    3.4641
    <u>.0424</u>   (Note that we rounded off to the nearest even integer)
    3.5065

Thus, your first intelligent guess at $\sqrt{12.3}$ is 3.5065. Squaring this value gives:

$$(3.5065)(3.5065) = 12.29554225 \simeq 12.3$$

So your first intelligent guess is quite close to the square root desired. If further accuracy is required, step 7 should be followed.

Note: *Most intelligent first guesses will be accurate enough for almost all statistical descriptive and inference purposes. A poor first guess, however, will mean further guesses and calculations.*

Example 2:

Find $\sqrt{.625}$.

    Using technique A we find that

$$\sqrt{.625} = \sqrt{\frac{(.625)(100)}{100}} = \frac{\sqrt{62.5}}{\sqrt{100}} = \frac{\sqrt{62.5}}{10}$$

So the problem boils down to one of finding $\sqrt{62.5}$, then dividing this answer by 10. (Note: even though 625 is a perfect square, 62.5 is not.)

Since 62.5 is halfway between 62 and 63, a first guess is that $\sqrt{62.5}$ is halfway between 62 and 63. Now, consulting the table we find

$$\sqrt{62} = 7.8740$$
$$\sqrt{63} = 7.9373$$

The difference is .0633 and half of this is $\frac{.0633}{2}$ = .0317

Adding .0317 to 7.8740 (or subtracting it from 7.9373) gives 7.9057 which is your first guess at $\sqrt{62.5}$.

Squaring 7.9057 gives 62.50009249 which is reasonably close to 62.5. Now dividing 7.9057 by 10 gives .79057 as the answer to $\sqrt{.625}$.

Exercises:
1. Find $\sqrt{33.7}$
2. Find $\sqrt{.084}$

Answers:
1. 5.8051 (1st guess)
2. .2897 (1st guess)

Appendix C

PROBABILITY*

Almost everyone has some notion concerning probability (likelihood, chance). Some of these notions are intuitive and some are based on familiarity with formal axiomatic systems. For our purpose here, however, we are concerned with three basic approaches to probability.

Subjective probability.

This approach refers to a belief, guess, hunch or estimate concerning an event and may be based on experience or intuition. For example, "It'll probably rain today," is a subjective probability statement. Also, "I'll bet it rains today" or "I think it'll rain today" are similar statements reflecting subjective probability. Subjective probability statements are rarely based on long-range observations and as a result, may be poor indicators of what really will happen.

Empirical probability.

Empirical probability, as the name implies, is based on an extended series of observation of an event. Here probability estimates may be quite accurate in reflecting what will happen. For example, insurance companies amass large amounts of empirical information prior to setting rates. Statements like, "the probability that a person will die prior to age 21 is .08" is an example of empirical probability since many observations were taken prior to making this observation.

The main difference between empirical and subjective probability is that empirical is based on extended observation and subjective does not have to be.

Empirical probability is defined as the relative frequency with which an event has occurred. For example, if a coin is tossed 100 times and 48 heads are observed, then an empirical estimate of the probability of "head" is 48/100 = .48. Note that all probabilities will be between 0 and 1 inclusive.

---

*Some of the terminology used herein is from *Psychological Statistics, Vol. II.* Individual Learning Systems, Inc. 1971.

Formal (axiomatic) probability.

Formal probability is the same as empirical probability with one exception. The counting for formal probability is done on a theoretical basis rather than an actual count, as in empirical probability. For example, instead of flipping a coin 100 times as above one could say, "If I assume the coin is fair and *if* I did flip the coin a large number of times, it would come up heads about 1/2 the time. Thus .5 would be the probability of a head." This is a formal probability statement. Note that the event, "coin stands on edge," would be given a formal probability value of zero since "heads" and "tails" is your only concern and you choose to disallow the event "stand on edge" even if it could occur. Since formal probability is the approach most commonly used and the one of primary interest in statistics another example is in order.

Suppose you have a jar with 10 red and 5 black marbles. What is the formal probability of drawing a black marble from the jar? Without even drawing a marble you know that 5 of the 15 marbles are black and you reason that if you did repeatedly draw a marble from the jar, replacing it everytime, you would get a black one 5 of 15 times. Thus the formal probability of a black marble on a single draw is

$$5/15 = 1/3 = .33.$$

The following are some basic laws and definitions of formal probability and a few will be stated here without proof. You should consult a basic probability text for further elaboration (The notation "P(A)" will represent the statement, "probability of event A.")

Definition 1: Let A represent any event. Also let ∅ represent "no event" and E represent the entire set of possible events. Then,

$$0 \leq P(A) \leq 1$$
$$P(\emptyset) = 0$$
$$P(E) = 1$$

Example:

Let the situation be a coin toss. There are two possible events allowed— "head" or "tails." Let A = head and B = tail. Now $P(A) = 1/2$, $P(B) = 1/2$ (both between 0 and 1). Let ∅ be the event "stands on edge." Since we have not allowed this event, $P(\emptyset) = 0$. Likewise, P (entire set of events) — P (head or a tail) — $P(E) = 1$, since a head or a tail *has* to occur.

Definition 2: If two events A and B cannot both occur at the same time, then A and B are said to be *mutually exclusive* and P(A and B) = 0.

Example:

P (head *and* a tail on coin toss) = 0, since *both* a head *and* a tail cannot occur on a single toss.

Definition 3: If two events A and B cannot affect or relate to each other in any way, they are said to be *independent* and P(A and B) = P(A) · P(B).

Example:

Let A = head on 1st toss of fair coin.
B = tail on 2nd toss of fair coin.

Since "head on first toss" and "tail on second toss" are not related in any way, then P(A and B) = P(A) · P(B) = (1/2)(1/2) = 1/4.

Law 1: P(A or B) = P(A) + P(B) − P(A and B)

Example:

Let A represent the event, "draw a king from a normal deck of cards." Let B represent the event, "draw a red card." Now law 1 gives

P(king or a red card) = P(king) + P(red card) − P(king and red card)
= (4/52) + (26/52) − (2/52)

since there are 4 kings, 26 red cards and two red kings in a normal 52 card deck. Thus:

P(king or a red card) = 28/52

Law 2: If A and B are mutually exclusive, P(A or B) = P(A) + P(B).

Note: By definition 2, P(A and B) = 0 since A and B are mutually exclusive. Thus Law 1 becomes P(A or B) = P(A) + P(B) − 0 which is Law 2.

Example:

Let A represent the event "draw an ace" and B represent the event "draw a king." On a single draw one cannot get both so A and B are mutually exclusive. Therefore,

P(ace or king) = P(ace) + P(king) = 4/52 + 4/52 = 8/52.

Law 3: If events A and A' comprise the entire set of possible events (A and A' are called complementary events) then,

P(A) = 1 − P(A')

Example:

Let A represent head on a single toss of a coin and let B represent a tail. Since these are the only two events allowed, A and B are complementary and

P(head) = 1 − P(tail)
       = 1 − 1/2 = 1/2

Definition 4: The conditional probability of event A given another event, B, is expressed by P(A/B).

Law 4: $P(A|B) = \dfrac{P(A \text{ and } B)}{P(B)}$

Example:

Let A be the event "draw an ace" and B be the event "draw a heart" then

$P(\text{ace}|\text{heart}) = \dfrac{P(\text{ace and heart})}{P(\text{heart})} = \dfrac{1}{52} \div \dfrac{1}{4} = \dfrac{4}{52} = \dfrac{1}{13}$

This is reasonable since the probability of an ace given you have to deal only in hearts is 1/13.

Appendix D

NORMAL PROBABILITY DISTRIBUTION

The normal probability distribution, sometimes called a probability density function, is very important in the behavioral sciences. Not only are its mathematical properties necessary for making statistical inference, but it describes a good many natural behavior phenomena. The distribution itself is continuous, symmetrical and bell-shaped with infinite range. Like most distributions, it has a mean, μ and standard deviation, σ. The mathematical formula for finding the height (Y) of the normal curve at any point (X) under the curve is

$$Y = \frac{e^{-1/2(X - \mu)^2/\sigma^2}}{\sigma \sqrt{2\pi}}$$

Our main concern, however, is not the curve itself, but the area or probability under the curve between, above and below specific points. Since the curve is a probability distribution the sum of all probabilities (areas) under the curve will be 1. This being the case, any specific part of the area under the curve will be a number between 0 and 1.

Some of the properties of the normal distribution are listed below: (see Figure 1 on page 105.)

A. Total area equal 1.
B. Symmetrical about the point $X = \mu$ (i.e., half the area above and half below the mean.) Note: mean = median = mode for a normal curve.
C. Actual range: from minus infinity to plus infinity.
D. Practical range: 3 standard deviations above and below the mean will contain almost all (99.74%) of the area.

Two standard deviations above and below the mean will contain approximately 95.44% of the area under the curve.
One standard deviation above and below the mean will contain approximately 68.26% of the area under the curve.

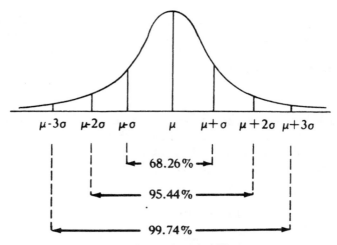

Figure 1. Normal probability curve.

It is apparent that when μ and σ change in value completely new normal distributions are produced. However, the areas will remain the same as well as the general symmetrical shape. Since all normal curves are the same as far as area is concerned, it will be handy to have one normal curve which would suffice for finding various probabilities, regardless of the mean and standard deviations used. There is such a normal and it is called the *standard normal* probability distribution. It is found by simply transforming the normal values (X) to standard normal values (z). The relationship between the X and the z is:

$$z = (X - \mu)/\sigma,$$

where μ and σ are the mean and standard deviation respectively of any normal curve. This transformation makes the mean of z distribution 0 and the standard deviation 1. The value of this transformation is that all normal curves can be reduced to this simple form. Figure 2 below is a graph of the standard normal curve. Area-wise it is identical to the curve in Figure 1.

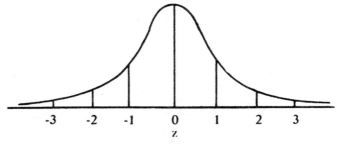

Figure 2. Standard normal curve.

Areas associated with several values of z have been tabled thus allowing us to approximate probabilities relative to any normal curve simply by transforming to the standard normal. Table D, Page 75, is such a table. Some typical examples will be presented below.

Example 1:

Suppose you could assume that the scores on a spatial perception test were normally distributed with a population mean of 78 and a standard deviation of 20. What proportion of scores should be less than or equal to 88?

In symbol form you are asking to find

$P(X \leq 88)$,

where X represents the spatial perception variable. Figure 3 shows the area of interest.

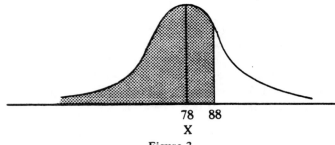

Figure 3.

Making the z score transformations and the score X = 88 becomes a z score expressed as,

$z_{88} = (88 - 78)/20 = 10/20 = .50$

The complete set of X scores transformed into z scores and the area of interest is shown in Figure 4.

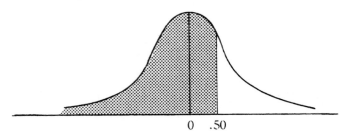

Figure 4.

Table D has standard normal z values along with the areas below (to the left) of these z values. Note that the z values start at 0 and go up to 3.49. Negative z values have to be converted (see next example) using the symmetry of the normal curve.

For this example z = .50 is in the table and since we wish to find

$$P(z \leqslant .50),$$

we read down the z column until we find z = .5. In the first column we read the value .6915. This is the probability of being less than or equal to a z value of .50. Thus,

$$P(z \leqslant .50) = P(X \leqslant 88) = .6915.$$

Example 2:

Assume the same situation as in Example 1 and find the probability (proportion) of a score being less than 68. That is, find

$$P(X < 68)$$

(Note that since the normal curve is continuous, this will be the same area as P(X ≤ 68)).

Figure 5 shows the area (probability) of interest.

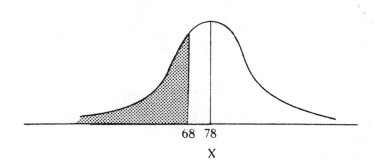

Figure 5.

The standard normal curve showing the same area is given by Figure 6.

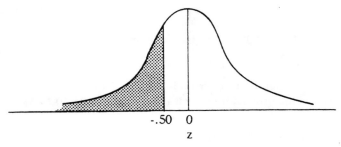

Figure 6.

Thus the probability is the area under the standard normal curve below z = −.50. This value is not in Table D but since the curve is symmetrical we can find the area below z = .50 and subtract from 1 to find the area below z = −.50 (see Figure 7).

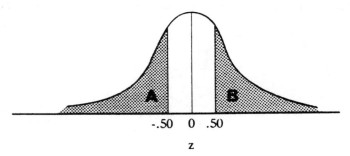

Figure 7.

Area A is the desired area and area B is the same area due to the symmetry of the normal curve.

From Example 1 we found that

$P(z \leq .50) = .6915$.

Thus, $P(z \leq -.50) = 1 - P(z \leq .50) = 1.000 - .6915 = .3085$. Therefore, $P(X \leq 68) = .3085$.

Example 3:

Find the probability that a spatial perception score is between 52 and 83, assuming the same distribution as in Example 1. This problem is the same as finding

$P(52 \leq X \leq 83)$. Figure 8 shows the desired probability.

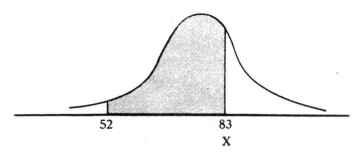

Figure 8.

Here the problem amounts to two problems with several different acceptable approaches. One such approach will be presented here. The steps are:
1. Find the area below X = 83 as in Example 1, i.e., find $P(X \leq 83)$
2. Find the area below X = 52 as in Example 2, i.e., find $P(X \leq 52)$
3. Find the difference between the 2 areas, i.e., find $P(X \leq 83) - P(X \leq 52)$ which will be the desired probability.

The steps in numerical form are:
1. $P(X \leq 83) = P(z \leq (83-78)/20) = P(z \leq .25) = .5987$
2. $P(X \leq 52) = P(z \leq (52-78)/20) = P(z \leq -1.3) = 1 - .9032 = .0968$
3. $P(52 \leq X \leq 83) = .5987 - .0968 = .5019$

Example 4:

Using the same distribution of Example 1, find the score such that 30% of the scores are below it, i.e., find the thirtieth percentile score.

Here the problem is reversed from the other examples, in that we wish to find a score associated with a given probability. Let us call the score X_{30} to denote the score such that 30% of the scores are below it. Now we must find X X_{30} so that

$$P(X \leq X_{30}) = .30$$

The desired area is shown in Figure 9.

109

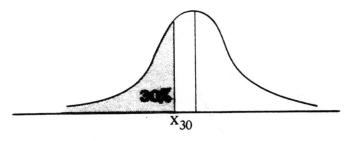

Figure 9.

Transforming to standard normal form the problem amounts to finding z, so that

$$P(z \leq (X_{30} - 78)/20 = P(z \leq z_{30}) = .30$$

Since Table D starts at an area (probability) of .5000 the value .30 cannot be found in the Table. However, knowing that the normal curve is symmetric, we need only find the negative of the z value corresponding to an area of .7000 in Table D. This value is approximately .52. Since our point is the negative of this, $z_{30} = -.52$ (See Figure 10)

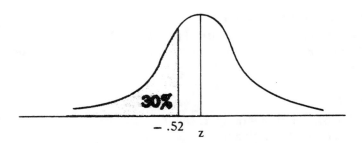

Figure 10.

Now that the z value is known we can find X because

$$z_{30} = (X_{30} - 78)/20 = -.52.$$

Solving for X_{30} and we find $(-.52)(20) + 78 = X_{30}$

$$67.6 = X_{30}.$$

Thus 67.6 is the 30th percentile.

Exercise:

For the Wechsler Intelligence scale for children (WISC), $\mu = 100$ and $\sigma = 15$. Assume these IQ scores are normally distributed and calculate the following:

    a.   The proportion of children with IQ scores between 90 and 115.
    b.   The score which is so high that only 1% of the children have scores above it. (That is, find the 99% percentile score.)

Answers:

    a. $z_{90} = .67$, $z_{115} = 1.00$.
           $P(90 \leq IQ \leq 115) = .5899$
    b.   I.Q. = 134.95

Appendix E

EXAMPLE CALCULATING FORMULA FOR MINIMUM SAMPLE SIZE

Consider the hypothesis testing situation

$H_o: \mu_1 = \mu_2$
$H_a: \mu_1 \neq \mu_2$,

where the assumptions of normality, homogeneity-of-variance and interval scale are made.

An approximation formula for the minimal group sample size where $n_1 = n_2 = n$ is given by

$$n = 2[(Z_{\alpha/2} + Z_\beta)\sigma/ES]^2$$

where $Z_{\alpha/2}$ and Z_β are the standard normal deviate values associated with $\alpha/2$ and β (1 − power) respectively, σ is the common standard deviation for both populations of scores and ES is effect size. (Note: $\alpha/2$ is used since the test is two-tailed. If the test were one-tailed then α would be used.)

Example:

Suppose you wished to test the above hypothesis and had *a priori* decided that ES = $1/4\sigma$, α = .05 and power = .90 (β = .10).

The minimal sample size for each of the samples now becomes

$$\begin{aligned}
n &= 2[(1.96 + 1.28)\ \sigma/.25\sigma]^2 \\
&= 2[(3.24)1/.25]^2 \\
&= 2[(3.24)(4)]^2 \\
&= 2[12.96]^2 \\
&= 2[169] = 338.
\end{aligned}$$

This means a *total* sample size of 676 is required to conduct the study with the preset α, β and ES.

The above formula is applicable only for the two-sample hypothesis given and should not be used with any other hypothesis set. Each hypothesis test has its own formula and may be quite different from the above.

Appendix F

MYTHS AND MISCONCEPTIONS IN BEHAVIORAL STATISTICAL METHODOLOGY

In the statistical methodology field, literally scores of texts and thousands of articles have been written by psychologists, educators, sociologists and various other behavioral scientists. These works, published largely by reputable publishing houses, probably represent the largest single collection of misconceptions, notions, biases, theoretical half-truths and errors known to the statistical world. One cannot help but wonder who reviewed these works of statistical magic-making before they were unleashed on an unsuspecting and naive public and before they mercifully began collecting dust on the shelves of professors who received complimentary copies. If this final and just resting place were their only place, then these literary efforts could be written off as harmless but, alas, some of the more horrendous textbooks actually have become "best sellers" and are thereby imposed on thousands of behavioral science students prior to their demise (the book or the student, whichever comes first). The myths and misconceptions which are most common in these widely used texts will be the focus of this appendix, for until they are eradicated, the behavioral sciences will continue to see a published parade of ill-conceived, poorly designed and inadequately analyzed studies masquerading as research.

The set of myths and misconceptions (hereafter referred to as M & M's) entertained herein is by no means complete but represents a collection sufficient to make the reader cast a jaundiced eye on the present-day textbooks and published articles in the behavioral sciences. Specific textbooks and other sources will not be cited since the reader will recognize almost all the M & M's as (1) something he has believed all his statistical life or, (2) something which seemed odd but since it was in print was taken on faith or, (3) something which seems so technical that he did not dare challenge it or (4) something he did not give a damn about anyway and therefore was trivial or (5) something known to the reader to be wrong per se but was classed as one of those "you know what I mean" statements (often accompanied by a literary wink of the eye).

The M & M's will be presented in forms most commonly found in the literature. Some are exact quotes and some are paraphrases, but the meaning will generally be unambiguous. With each statement will be a brief discussion of the reasons the statement is an M & M, followed by conjectures as to the effect of this M & M on subsequent understanding of statistical methodology and applications in the behavioral sciences. The order in which the statements are presented connotes absolutely nothing and matters very little since an M & M is an M & M, no matter when it occurs.

Some M & M's

1. Alpha and beta are inversely related.

This statement is true *only* if effect size (ES) and sample size (n) are fixed. Since alpha and beta are *a priori* values of judgment and are, along with ES, the determinants of the sample size, it is hard to believe that the statement is still found in the literature. For example, both alpha and beta could be preset at .001 if the researcher so desired. Most texts wherein the statement appears make the tacit and unrealistic assumption that ES and sample size are fixed but do not bother to inform the reader of this assumption.

Belief in this M & M could keep a researcher from designing a study where both alpha and beta were small, whereas any serious researcher would admit that to have low error rates is an ideal state. This belief that one must give up one kind of error to have another is contrary to the goal of any good piece of research, i.e., to minimize both error rates. What the researcher must give up if he wants very small alpha and beta values is very small effect sizes or small sample sizes or both.

2. Alpha and p, the probability of the data given H_o is true, are the same.

Here the error is in equating a subjective preset value, namely alpha, with a calculated probability which is independent of alpha, namely p. The fact that they may be equal is the result of a numerical happenstance, not a definitional equivalence.

The researcher who believes this M & M and who correctly presets alpha at .05, say, cannot help but be amazed when the calculated p is .023, for example. Alpha does not now automatically become .023 since this would destroy the preestablished balance between alpha, beta, effect size and n. The only time alpha and p are discussed together is when the researcher wishes to make a statistical decision; i.e., if p is less than or equal to alpha, then reject H_o; otherwise, do not reject.

3. The statistical test approached significance.

This reflects the ultimate attempt by the researcher to squeeze statistics dry in the face of a failure to reject H_o at some preset alpha level. This statement is pure nonsense, since a statistical test rejects (i.e., is significant) or not at a preset alpha level. A single test result cannot "approach" anything. Taken to its extreme, this M & M would imply that any test result "approached" significance, since any test "almost" rejects H_o if "almost" is loosely defined. (The same misconception is responsible for statements like "highly significant," "tends to be significant," etc.)

4. The t-test is a small sample statistic.

The use or derivation of the t-test has nothing to do with whether the sample size is large or small. Basically, if the population variances are known, then a z-test should be used (if the other assumptions hold). If the variances are unknown, then a t-test is appropriate. This M & M probably is an attempt to say, "the normal curve is a good approximation of the t-distribution if sample size is large but not so good for small samples." What is baffling to this writer is why an author would go out of his way to make this misleading approximation statement when no statement is needed.

A researcher could be led by this M & M to believe that if a sample size is very small, a t-test *should* be conducted when no test, including the t-test, could be defended due to the possible low power.

5. The chi-square test assumes normality.

This bit of mythology probably comes about as a result of a researcher being exposed to a chi-square derivation which invoked the Central Limit Theorem. But neither the use of the test nor its derivation assumes normality. To invoke the Central Limit Theorem for a derivation is not making an assumption of normality; no more than proving the Central Limit Theorem itself makes the assumption of normality.

A belief in this M & M might influence the researcher to make unnecessary and unrealistic assumptions for situations suited to a chi-square test.

6. A sample can be so large that trivial differences will be detected.

In this statement the misconception is in equating statistical "significance" with nontrivial or important. The serious researcher would have stated prior to the sample collection what size true differences were important to detect via an effect size (ES) level. This would be his expression of what true differences were trivial and what differences were not. The subsequent result of the statistical test cannot then tell the researcher what is important, only that the probability of his data given H_o is smaller than alpha or not. If, after collecting his sample (based on preset alpha, beta and ES), the researcher observes differences so small that he considers them trivial, then no statistical test should be conducted because, regardless of the statistical test outcome, the differences are still trivial by his judgment. If a researcher, after considering alpha, beta, and effect size, finds he needs a sample size of 100 and can obtain, with no real strain, a sample of

1000, he should do so. The larger the sample, the better, if for nothing other than making better parameter estimates. This larger-than-needed sample size will not automatically detect trivial differences, since triviality has been prejudged by the researcher, but it will allow the researcher to reduce alpha and beta. Note that the researcher's view of what is trivial is not affected by how many observations he takes. If, for example, the researcher felt that .2 was an important true difference and anything smaller was trivial, then it matters not whether the study is conducted with 10 or 10,000 observations. Any difference below .2 is still trivial regardless of a statistical test or sample size. (This is not to say, however, that a researcher's view as to what is important cannot be affected by repeated observations of real data.) This M & M could cause a naive researcher to throw away a lot of good random data for fear that with too much data he will find smaller trivial differences. There is no statistical justification for discarding good data.

7. Parametric tests are more powerful than nonparametric tests.

This M & M probably came about as a result of parametric zealots trying to justify the applications of a t-test, for example, under all conditions and situations. The statement is true, however, only if all the parametric assumptions are met. If they are not met, then in most cases power cannot be calculated, much less said to be larger than something else. Bradley (1968), Blair and Higgins (1980) and others have stated that if the parametric assumptions do not hold, then some nonparametric tests are actually more powerful than the analogous parametric test. This seems reasonable when considering a nonparametric test with power efficiency of 96 percent, say. If the parametric assumptions are met, then only a 4 percent boost in sample size is needed for the nonparametric test to equal the parametric test in power. If the parametric assumptions are not met, then it is reasonable that the test whose assumptions are met should be the more powerful, since the two tests were comparable in power under the most restrictive of conditions.

The danger of this M & M is that a researcher could be influenced to disregard alternative procedures (both parametric and nonparametric) for his study, thereby denying himself and his readers an exposure to a broader spectrum of admissible, defensible techniques. Nothing but good can come from being well-informed about alternate approaches and assumptions.

8. A simple random sample can be obtained by assuring that each observation has an equal chance of being selected.

Although not incorrect per se, this M & M is the *consequence* of a random sample, not its definition. In its most basic form a simple random sample of size n is one so selected that every *sample* of size n will have an equal chance of being selected (see Cochran [1963] and any other sampling theory text).

An example will demonstrate why this definition is not the same as assuring that each observation has an equal chance of being selected.

Consider the front row of a classroom consisting of 10 students. Suppose you wish to obtain a simple random sample composed of 5 of these students. To assure that each student has an equal chance, you could flip a fair coin and if "heads," then start with the person on the end of the row and take every other person. If "tails," start with the second person and take every other person. The probability of any one student being in the sample is 1/2, but have you assured that every sample of size 5 had an equal chance? Obviously not, since no two consecutive people would be in the sample, or the last five, etc.

If the researcher typically assigns every observation a number, and picks numbers at random using a random number table or random number generator, then a simple random sample is assured, not because the observations have an equal chance but because each sample does. (Obviously, if each sample has an equal chance, so does each observation.)

Taking this M & M as a definition could cause researchers to select samples biased to some degree, with the nature of the bias being difficult to detect.

9. *The larger the sample size, the more normal the distribution becomes.*

The essence of this M & M is the belief that the assumption of normality required for parametric tests is better met if the sample size is larger. This is a misconception in that a true distribution of scores does not get more normal simply by taking larger samples. Suppose the true distribution of scores is bimodal; then by taking more and more observations the researcher should become increasingly aware that he is dealing with a non-normal distribution.

The origin of this M & M probably stems from the Central Limit Theorem, whereby normality is approached when larger and larger numbers of *averages* of scores are obtained. The fact that averages look normally distributed does not, of course, mean that the original scores were normal, since averages of almost any kind of numbers will tend to be normal by the Central Limit Theorem.

A belief in this M & M could cause a researcher to inappropriately apply a parametric technique. He/she would mistakenly gain some level of comfort relative to the normality assumption simply by thinking his/her sample was "large enough."

10. *The value* $S = \sqrt{\frac{\Sigma(X - \bar{X})^2}{n - 1}}$ *is an unbiased estimate of* σ.

This common myth results from erroneous algebraic juggling of the definition of expected values, since it is true that the expected value of S^2 is σ^2. It is, however, not true that the expected value of S is σ. Thus S is a biased estimator of σ.

One may well ask, "What difference does it make?" But the one who asks this evidently does not care for correct concepts, only a biased convenience. The obvious answer to the need for correct statements of unbiasedness is that the basis for use of the t-test and F-test and almost all classical statistics tests rests on having obtained unbiased estimates of σ^2, not unbiased estimates of σ. One takes the square root of a variance for usage only, not so that the theory holds—variances are the keys to inference theory, not standard deviations.

REFERENCES

Bradley, James V. *Distribution-Free Statistical Tests.* Englewood Cliffs, N.J.: Prentice-Hall, 1968.

Brewer, James K. *Introductory Statistics for Researchers.* Edina, MN: Bellwether Press, 1986.

Blair, C; and Higgins, J. A comparison of the power of Wilcoxon's Rank-Sum statistic to that of Student's t-statistic. *Journal of Educational Statistics,* vol. 5, No. 4, Winter 1980. 309–335.

Cochran, William G. *Sampling Techniques.* New York: John Wiley, 1963.

Cohen, J. *Statistical Power Analyses for the Behavioral Sciences.* New York: Academic Press, 1969.

Davies, O.L. *Statistical Methods in Research and Production.* New York: Hafner Publishing Co., 1961, Chapter 5.

Dixon, W.F. and Massey, F.J., Jr. *Introduction to Statistical Analysis.* (2nd Ed.). New York: McGraw-Hill, 1957.

Glass, Gene V., Peckham, P.D. and Sanders, J.R. Consequences of failure to meet assumptions underlying the analysis of variance and covariance. *Review of Educational Research,* 1972, Vol. 42, No. 3, 237-288.

Guenther, William C. Power and sample size for approximate χ^2 test. *The American Statistician,* 1977, 31, 2, 83-85.

Tversky, Amos and Kahnman, Daniel, Belief in the law of small numbers. *Psychological Bulletin,* 1971, 76, 105-110.

INDEX

A

academic crook, 55
alpha (α), 41
alternate hypothesis, 39
analogous H_o, 37
arithmetic mean, 6
association,
 measure of, 20
assumptions for,
 $H_o: \mu$ = constant, 47
 $H_o: \mu_1 = \mu_2$, 52
 $H_o: \sigma_1^2 = \sigma_2^2$, 56
 $H_o: \rho = 0$, 58
 chi-square test, 60
average, 6

B

basic mathematics skills refresher, Appendix A
Bayesian statistics, 63
bell-shaped curve, Appendix D
best linear equation, 25
beta (β), 43
biomodal, 18

C

calculating formulas for
 S^2, 13
 r, 22
causality, 30
central location, 6
 measure of, 6
chi-square, 60
 table value, χ^2, Table C
classical inferential statistics, 64
coding, 14

confidence intervals, 50, 51
 on μ, 50
 relation to hypothesis testing, 51
continuous variable, 2
correlation
 meaning of, 20
 measure of, 20, 22

D

data, 1
data description, 1
definitional formulas for
 S^2, 11
 $\bar{X}$, 6
 r, 22
degrees of freedom for
 $H_o: \mu$ = constant, 47
 $H_o: \mu_1 = \mu_2$, 52
 $H_o: \sigma_1^2 = \sigma_2^2$, 56
density, 15
dependent variable, 28
descriptive statistics, 2
deviation, 12
discrete variable, 2
dispersion, 11
distribution, 7
distribution-free, 36

E

effect size (ES), 41
empirical probability, Appendix C
error of measurement, 3
error types, 41, 43
estimate,
 of μ, 6
 of σ^2, 11

of ρ, 22
unbiased, 12
examples
  for testing
    $H_o: \mu$ = constant, 48
    $H_o: \mu_1 = \mu_2$, 52
    $H_o: \sigma_1^2 = \sigma_2^2$, 57
    $H_o: \rho = 0$, 58
  of confidence intervals, 34
exercises with
  $\overline{X}$, 6
  S^2, 11
  confidence intervals, 34
  correlation, 22
  expected values, 60
  regression, 27
  standard error of estimate, 28
  testing $H_o: \mu$ = constant, 49, 50
  testing $H_o: \mu_1 = \mu_2$, 54, 55
  testing $H_o: \sigma_1^2 = \sigma_2^2$, 56, 57
  testing $H_o: \rho = 0$, 59
  testing H_o: X and Y are independent, 60
  confidence intervals, 51
expected frequencies, 60
expected value, 60
experimenter inference, 35

F

F_α table values, Table B
F-ratio, 56
formal probability, Appendix C
formulas for
  $\overline{X}$, 6
  S^2, 11
  r, 22
  confidence intervals, 51
  slope and intercept, 27
  standard error of estimate, 28
frequency
  distribution, 15
  curve, 15
  histogram, 15

G

general relationships between, α, β, ES, and n, 41

H

H_o, 39
histogram, 15
homogeneity of variance, 57
hypotheses, 39
hypothesis test (examples)
  $H_o: \mu$ = constant, 48
  $H_o: \mu_1 = \mu_2$, 52
  $H_o: \sigma_1^2 = \sigma_2^2$, 57
  $H_o: \rho = 0$, 58
  H_o: X and Y are independent, 63
hypothesis testing
  general concepts, 34, 40
  hypothesis-specific concepts, 40

I

independent samples, 52
independent variables, 28
inference from samples, 35
inferential statistics, 2
inferential tests, 35
interval measurement, 4
intra-sample independence, 52

L

least squares, 26
likelihood, 44
  under $H_o: \mu$ = constant, 46
  under $H_o: \mu_1 = \mu_2$, 52
linear relationship, 21

M

math skills refresher, Appendix A
maximum rationalization, 64
mean, arithmetic
  exercise with, 6
meaning of
  correlation, 20
  r, regression and S.E.E., 28
measurement, 3
  error, 3
  interval, 4
  nominal, 3
  ordinal, 4
  ratio, 4

measures of
   association, 20
   central location, 6
   dispersion, 11
   predictability, 28
median, 10
mode, 10

N

nominal measurement, 3
nonparametric statistics, 36
nontrivial differences, 41
normal, Appendix D
normal table values, Table D
notation, 67, 68
null hypothesis, 39

O

objectives for basic math skills,
   Appendix A
observation, 3
observed frequencies, 60
one-tailed test, 40
ordinal measurement, 4

P

parameter, 3
parametric statistics, 36
Pearson Product Moment Correlation
   Coefficient, 20
plausible ES values, 42
population, 1
   correlation, 20
   mean, 6
   median, 10
   standard deviation, 12
   variance, 11
power, 41
predictability, 28
prediction, 25
probability, 2
   empirical, Appendix C
   formal, Appendix C
   subjective, Appendix C
purpose of
   correlation, 20
   sample, 3
   statistics, 1

Q

Questions concerning
   association, 21-33
   assumptions of hypothesis tests, 36
   calculating formulas, 13
   central location, 6-10
   confidence intervals, 51
   degrees of freedom, 49
   dispersion, 11-14
   effect size, 41
   inference, 34-42
   likelihood, 44
   one-tailed tests, 50
   parametric vs. nonparametric, 36
   prediction, 25
   random samples, 35
   regression, 25
   robustness, 38
   sample size, 36
   Type I and Type II error, 41, 43
   unbiased estimators, 12

R

random sample, 35
range, 11
ranked data, 30
ratio measurement, 4
references, 119
regression, 25
rejection, 25
related samples, 52
relationships between α, β, ES, and
   n, 41
robustness, 38
rules on $\bar{X}$ and S^2, 14

S

Sample, 1
   correlation, 22
   median, 10
   mode, 10
   size, 36
   standard deviation, 12
   variance, 11
sampling distribution, 44
scatter plots, 23
significant, 42
simple random sample, 35

skew, 18
Spearman Rho, 20
standard deviation, 12
standard error of estimate, 28
    exercise with, 28
standard scores, 15
statistical inference, 35
statistical reminders, 66
statistics, 3
    purpose of, 1
student's t-test, 47
    table values t_α, Table A
subjective probability, Appendix C
summation notation, Appendix A
symmetric, 17

T

terms and notation, 67, 68
test statistic for
    $H_o: \mu = $ constant, 47
    $H_o: \mu_1 = \mu_2$, 52
    $H_o: \sigma_1^2 = \sigma_2^2$, 56
    $H_o: \rho = 0$, 58
    $H_o:$ X and Y are independent, 60
testable hypothesis, 39

transformation, 14
two-tailed tests, 40
Type I error, 41
Type II error, 43

U

unbiased, 12
uniform, 18
unimodal, 18

V

valid experimenter inferences, 35
variable, 2
variance, 11

X

X on Y line, 25

Y

Y on X line, 25

Z

zero correlation, 23